D0142351

THE WHITE PLANET

THE WHITE PLANET

THE EVOLUTION AND FUTURE
OF OUR FROZEN WORLD

Jean Jouzel, Claude Lorius,
and Dominique Raynaud

Translated from the French by Teresa Lavender Fagan

PRINCETON UNIVERSITY PRESS

Princeton and Oxford

First published in France under the title *Planète blanche, les glaces,*
le climat et l'environnement © Odile Jacob, 2008.

English translation, adapted and revised, copyright © 2013 by Princeton University Press
Requests for permission to reproduce material from this work should be sent to
Permissions, Princeton University Press
Published by Princeton University Press, 41 William Street, Princeton, New Jersey 08540
In the United Kingdom: Princeton University Press, 6 Oxford Street,
Woodstock, Oxfordshire OX20 1TW

press.princeton.edu

Jacket Photograph: Fox Glacier, South Island, New Zealand. Te Wahipounamu UNESCO
World Heritage Area. © Cloudia Newland. Courtesy of Shutterstock.

Cartoons by Michel Creseveur

All Rights Reserved

LIBRARY OF CONGRESS CATALOGING-IN-PUBLICATION DATA
Jouzel, Jean, 1947-
[Planète blanche. English]
The white planet : the evolution and future of our frozen world / Jean Jouzel, Claude Lorius,
and Dominique Raynaud ; translated from the French by Teresa Lavender Fagan.
p. cm.
"First published in France under the title Planète blanche, les glaces,
le climat et l'environnement, Odile Jacob, 2008."
Includes bibliographical references and index.
ISBN-13: 978-0-691-14499-3 (cloth : alk. paper)
ISBN-10: 0-691-14499-0 (cloth : alk. paper) 1. Glaciers. 2. Glaciology. 3. Climatic changes.
4. Paleoclimatology. 5. Greenhouse effect, Atmospheric. I. Lorius, Claude.
II. Raynaud, Dominique. III. Title.
QC981.8.I23J6813 2013
551.31--dc23   2012028432

British Library Cataloging-in-Publication Data is available

Ouvrage publie avec le concours du Ministere francais charge de la culture—
Centre national du livre.
This book has been published with support from
the French Ministry of Culture / National Book Center.

This book has been composed in Garamond Premier Pro

This book is printed on recycled paper ♻

Printed in the United States of America

1 3 5 7 9 10 8 6 4 2

To Jean-Marc Barnola,
glaciologist and humanist, who liked so much to visit our White Planet.

PHOTO COURTESY OF J. CHAPPELLAZ, CNRS-UJF-IPEV.

CONTENTS

PREFACE   XI

PART ONE   THE WORLD OF ICE: PAST AND PRESENT   1
    Chapter 1   The Ice on Our Planet   3
        Snow and Ice: A Multifaceted World   3
        Mountain Glaciers and Ice Caps   5
        Polar Regions: The Omnipresence of the White Planet   7
        Greenland, Antarctica, and Ice Shelves   10
        Ice: An Agent and Indicator of Climate Change   14
        The White Planet and Sea Levels   16

    Chapter 2   From Exploration to Scientific Observation   18
        The Flow of Mountain Glaciers   19
        Mass Balance: The Health of a Glacier   21
        The Arctic Ocean in the Time of the Explorers   23
        The Arctic Ocean: Vulnerable Ice   25
        Greenland: An Island Inhabited for Millennia   28
        Greenland: An Increasingly Negative Mass Balance   28
        Antarctica: A Much More Recent Exploration   31
        Antarctica: A Long Uncertain Mass Balance   34

    Chapter 3   Ice through the Ages   37
        The Time of the Pioneers   37
        Ice of Long Ago   40
        Glaciations of the Quaternary and Astronomic Theory   46

PART TWO   POLAR ICE: AMAZING ARCHIVES   51
    Chapter 4   Reconstructing the Climates of the Past   53
        The Round of Isotopes   54
        Going Back in Time   57
        The Recent Period   57
        The Distant Past   60
        Paleoceanography   61
        Continental Archives   62

Dating Oceanic and Continental Archives   64
A Cornucopia of Results   66

Chapter 5   Glacial Archives   68
The Long Story of a Snowflake   68
The Ice and Its Isotopes: A Paleothermometer   70
Impurities with Multiple Sources   71
Air Bubbles in the Ice: A Very Beautiful Story   73
The Headaches of Dating   77

Chapter 6   The Campaigns   82
Camps Century and Byrd: The First Deep Ice Core Drillings   82
Fifty Years Ago: The French on the Polar Ice   86
The First Drilling at Dôme C: Success of the French Team   88
Rapid Climate Variations: Initial Inklings   91
Vostok: A Collaboration between French and Soviet Teams   92
*Europe and the United States: Two Drilling Operations in the Center
   of Greenland*   96
Europe Turns to Antarctica   98
Vostok: More than 3,600 Meters of Ice   101
Other Core Drilling in Antarctica   103
The Glaciers of the Andes and the Himalaya   105
A Return to Greenland   106
The European EPICA Drilling: A Double Success beyond All Hopes   108

Chapter 7   Vostok: The Cornucopia   110
A Complete Glacial-Interglacial Cycle   112
Climate and Greenhouse Effect Go Hand in Hand   113
Much More Information   118
A Huge Lake under the Ice   120

Chapter 8   Dôme C: 800,000 Years and the Revolution of the
   Rhythm of Glaciations   122
Ice Older than That at Vostok   123
Inversion of the Magnetic Field   126

Chapter 9   Rapid Climatic Variations   130
The First Indications   130
Increasingly Clear Indications   132
A Connection with Ocean Circulation?   133

Confirmation 134

Rapid Events during a Warm Period? 139

Initially Underestimated Changes in Temperature 142

The Connection with the Ocean Henceforth Demonstrated 143

Consequences on a Planetary Scale 147

Chapter 10   The Last 10,000 Years: An Almost Stable Climate 149

Volcanism and Solar Activity: Natural Climatic Forcings 150

How Long Has Human Activity Been Changing the Composition of the Atmosphere? 152

PART THREE   THE WHITE PLANET TOMORROW 157

Chapter 11   The Climate and Greenhouse Gases 159

The Greenhouse Effect: A Truly Beneficial Natural Phenomenon 159

The Greenhouse Effect Due to Human Activity: A Slow Awareness 163

How Did We Get to This Point? 168

Chapter 12   Have Humans Already Changed the Climate? 173

The Time of the Pioneers 173

The Awareness 174

The Establishment of the IPCC 177

The Problem of Aerosols 180

The Climate in the Last Millennium 183

Warming Is a Certainty 185

The Arguments of Skeptics 189

The White Planet on the Front Lines of Global Warming 195

Chapter 13   What Will the Climate Be in the Future? 201

A True Upheaval if We Aren't Careful 202

What Will Become of Our Glaciers? 206

An Arctic Ocean without Ice? 209

Surprises under the Frozen Ground 210

A More Rapid and Higher Sea-Level Rise than Predicted 211

The Halt of the Gulf Stream 214

Chapter 14   A Warming with Multiple Consequences 218

A True Upheaval on a Global Scale 218

Mountain Regions 222

Polar Regions: Multiple and Diverse Impacts 223

The Political and Economic Stakes: Climate and Oil 225

Chapter 15   What We Must Do   227
  Stabilizing the Greenhouse Effect: A True Challenge   228
  The Kyoto Protocol: A First Step   230
  The Bali Conference   234
  Can the Challenge Be Met?   236
  Copenhagen: Failure or Half-Success   238
  A Necessary Adaptation   241
  The "Grenelle de l'environnement"   242

PART FOUR   THE POLES AND THE PLANET   245
Chapter 16   The Crucial Place of Research   247
  A Short History of the Polar Years   249
  The International Polar Year 2007–2009   251
  Glacial Ice Coring: Ambitious Objectives   253
  *The Microbiology of Ice and Subglacial Lakes: Life in an Extreme
    Environment*   255
  Concordia: A Station Full of Promise   258

Chapter 17   Humans and the Rise of Pollution   261
  The Story of Lead   262
  Other Heavy Metals, Including Copper   264
  Sulfates   266
  Radioactivity   268
  The Ozone Hole: An Emblematic Pollution   269
  The Anthropocene and Greenhouse Gases   271

CONCLUSION: THE ANTHROPOCENE ERA   272

NOTES   277

SELECTED BIBLIOGRAPHY   289

INDEX   291

PREFACE

Do you know why the land and ice in the surroundings of the North Pole are called Greenland—in French, Groenland? You might think that the name comes from an old *inuk* word, but Erik the Red supposedly named the island when he founded a Viking colony there in 984 A.D. Some historians claim that Erik the Red invented the term *green land* to entice his kinsmen to that desolate land. This is perhaps not really true, because even if Greenland seems today to be a huge white expanse, along certain fjords one can still see green fields where animals are raised. Between 984 and the fifteenth century this was "the most distant fore-posts of European civilization." "Scandinavians 1,500 miles from Norway built a cathedral and churches, wrote in Latin and Old Norse, wielded iron tools, herded farm animals, followed the latest European fashions in clothing—and finally vanished."[1] The stone church in Hvalsey endured; the Vikings of Greenland, who numbered five thousand in 1000, disappeared, while their neighbors the Inuits barely survived.

Around the year 800 Scandinavia was warming up, but the cultivatable lands in its mountainous regions and along its rocky coasts were too few to feed the large Viking population. On their fast-sailing ships that were capable of long voyages, the Vikings set off in search of more abundant lands. Some eventually settled under the Sicilian sun, but in the North Atlantic they founded several colonies: in the Orcades, the Shetland Islands, the Feroe Islands, in Iceland, and in Greenland. From there, the descendants of Erik the Red even tried to settle in a land they called Vinland, which today would include the coasts of Canada south of Labrador as well as Newfoundland, the Gulf of Saint Lawrence, and part of the coast of New England. But their attempts quickly failed, it seems, owing to a lack of means and men to fight the Indians. Those intrepid adventurers thus returned home to the shores of Greenland, which were more peaceful though less hospitable. Four hundred years after Erik the Red landed, only the ruins of farms where his countrymen attempted a life remained.

"The climate became too cold, and they began to die off," wrote an archae-
ologist. In fact, between the ninth and fourteenth centuries the climate could
have warmed to such an extent that when the Scandinavians reached Greenland around the year 1000 they found its climate a bit more propitious for
farming and raising livestock. But the cooling of the Little Ice Age, which
lasted until the nineteenth century, doesn't explain everything. It was above
all the Vikings' inability to adapt their way of life, their values, their social
structures, and their economy that caused them to die off. As Jared Diamond
explains, in a hostile environment, collapse is not inevitable; it depends on
the choices a society makes: "Environmental damage, climate change, loss of
friendly contacts with Norway, rise of hostile contacts with the Inuit, and the
political, economic, social, and cultural setting of the Greenland Norse.
Greenland provides us with our closest approximation to a controlled experiment in collapses."[2]

Today an even greater possibility of collapse threatens us because it involves the entire planet. We must turn to the icy lands to fully understand the
nature and magnitude of the threat. But it is no longer a question of the ability or inability of some to react adequately to a temporary and local change
in climate—our current situation is unlike that of the Mayan lords who were
exhausted by wars and unable to foresee the consequences for their people of
repeated droughts and overexploited land.

We are dealing with a perceivable degradation of living conditions in our
society that is exacerbating the crises that already exist: poverty, access to
clean water and to sources of energy, migrations, geopolitical destabilization
and conflicts. Subsequently, the Inuits are protesting; the polar bear, the largest living land carnivore, is threatened; and the krill, very useful little shrimp,
as well as the seals and the penguins in the southern seas that eat them, seem
at risk. More than a sixth of the world's population, most of which is in Asia,
lives in regions that rely on the water from snow and glaciers: if those shrink
in volume that could affect the future of those regions. One hardly dares
mention the consequences on tourism of the melting of the glaciers in the
Alps or the Pyrenees or of a shrinking of the snow cover that might be caused
by a "mere" increase of 2°C. Our vacations would be seriously affected—but
that would be the least of our problems.

"The Vikings were doomed from the beginning," Jared Diamond has written. And the petty Mayan kings were too concerned with their wars to sense

any danger. Will we be able to react? The 2007 Nobel Peace Prize attributed to, Al Gore and to the Intergovernmental Panel on Climate Change (IPCC) testifies that there is a growing awareness of the threat that climate change is posing to the planet. Glaciology and glaciologists have contributed to this heightened awareness.

More than fifty years ago, in 1957, one of the authors of this book, Claude Lorius, went to Adélie Land. His intention, within the framework of the International Geophysical Year, was to explore and learn more about Antarctica. Temperatures, the thickness of the ice, snow accumulation, the advance of the glaciers: everything was new, and there was much to learn. A few years later, around 1965, a new path was opened: ice analysis enabled glaciologists to determine the temperature of the atmosphere at the time the snow had fallen. These new data were a true gold mine, which has been developed in France in collaboration with the Commissariat à l'Energie Atomique (CEA) within which one of the authors of this book, Jean Jouzel, spent the greater part of his scientific career. At the Laboratoire de Glaciologie et de Géophysique de l'Environnement (LGGE) du CNRS (Centre National de la Recherche Scientifique) in Grenoble, we analyzed the impurities contained in the ice, in particular the air bubbles that are evidence of past atmosphere, a realm that our third author, Dominique Raynaud, brought to the forefront.

In 1987 we demonstrated that the variations in temperature and the amount of greenhouse gases in the atmosphere were connected throughout the last climatic cycle (in the last 150,000 years). With the work of American, Danish, and Swiss researchers, who focused more on Greenland, and the Franco-Russian collaboration centered on Antarctica, the study of glacial archives was well under way. We could henceforth attempt to reconstruct and to understand the mechanisms of climate variations in the past. And from that, it becomes possible to speculate about the future climate and its effects worldwide.

It is the development of this new discipline that we wish to reveal in this work. We are fortunate to have been able to contribute to its inception. But we owe a great deal to the researchers, technicians, administrators, and people on-site who have accompanied us during this amazing scientific adventure. We wish to thank them all.

These glacial archives are of interest not only to the scientific community. They are of huge interest to the broader public because the polar regions in particular remain synonymous with adventure. And at present everyone senses their fragility. Anyone who is interested in the climate, in its history, its future, cannot ignore these regions of extreme conditions.

Several recent films have used the climate as their central theme. One of the two we will mention is a documentary that has become emblematic of the "global warming" concern for the future: *An Inconvenient Truth*, of which Al Gore is the indisputable star. The other, *The Day after Tomorrow*, is a work of fiction that quickly departs from the realm of the believable. Both assign a large role to the results obtained from deep glacial core drilling. These two films placed a camera lens on the research undertaken in the polar regions, which have gained new life during the International Geophysical Year, whose fiftieth anniversary we celebrated during the International Polar Year 2007–2008.

We can reconstruct the climates of the past using various methods, and the ice of the ice sheets are a precious memory of our climate. But they above all contain a unique treasure with their air bubbles. As the climate has evolved these bubbles have recorded the composition of our atmosphere, including greenhouse gases. This is the central theme of our book: we will retrace the campaigns of exploration and of core drilling that have reached close to four thousand meters in depth and present the measurements and interpretations that they have allowed. But first we will familiarize the reader with what we scientists call the "cryosphere" and summarize what we know about the "ice of yesteryear," which has impacted the life of our planet from the time the Earth almost became a snowball up to the heat and the cold of the last two million years.

Our shared enthusiasm for the wealth of results offered by deep core drilling in Antarctica and Greenland comes from the fact that beyond what they teach us about the way in which our climate evolved in the past, they are rich in information on its future evolution. That is why this book also looks at the future while emphasizing the impact of climate warming on the white planet, particularly on the polar regions, which are particularly sensitive to it.

At the end of the journey, we will better understand why, since the beginning of the nineteenth century, we have entered a new era, the Anthropocene, which is characterized by an increase in pollution due to the activities of humankind, which puts its stamp on the environment and on the climate of our planet and thus on our future. This book does not propose a miraculous solution. Its object is to describe the progress of a science that gives us every reason to be concerned and aims to inspire citizens and policymakers to confront the enormous challenge before us.

Let us now look at the icy expanses that will first take us back in time.

Acknowledgments

This work presents scientific results that are the fruit of work carried out by many researchers, engineers, and technicians in areas relating to the history of our climate and of our environment, as well as their future evolution. We wish to thank everyone involved, with special thanks to those we have collaborated with, in France and abroad, during the many years that we have spent deciphering the glacial archives. It has been a fantastic collaborative adventure. In France, since the 1960s, we have enjoyed a close collaboration among the teams from Grenoble, Saclay, and Orsay, and that collaboration has benefited from the unfailing logistical support of the Expéditions polaires, then of the IPEV (Institut Polaire Français Paul Emile Victor), and from that of various departments of the CNRS and the CEA, as well as from European programs (European Union and the European Science Foundation). The adventure has unfolded within a context of remarkably organized and extremely stimulating international collaborations that were enhanced through many friendships. Our thanks go out to everyone. Thanks, too, to Volodia Lipenkov and to Jean-Robert Petit and Valérie Masson-Delmotte for advising on certain chapters in the book.

PART ONE

THE WORLD OF ICE: PAST AND PRESENT

The Ice on Our Planet

The temperature conditions that govern our planet are such that water can exist in three forms (vapor, liquid, and solid) in proportions that vary in different climate conditions. All of the water on the Earth makes up what is called the hydrosphere. Most water exists in liquid form, 97% of which is in the seas and oceans, which cover 360 million km², or more than 70% of the planet's surface. Freshwater, a vital resource found in the ground, lakes, rivers, and above all aquifers, represents only a small proportion of the total, a bit more than 0.5%; water vapor in the atmosphere amounts to .001%; and the water of the living world, the biosphere, represents much less.

Low temperatures are found in high altitudes and high latitudes; this is where ice forms. In current climate conditions, this ice represents only around 2% of the hydrosphere. How is that 2% distributed? How is it formed? What surface does it cover? What climatic role does it play? Before looking to the past, we will describe our white planet as it is now. Table 1.1 shows its main characteristics.

Snow and Ice: A Multifaceted World

When we think of ice, we immediately think of 0°C (or 32°F). This is the temperature at which water freezes and ice melts. It is a reference point for the temperature scale that we use in our everyday lives. In nature, it is a bit more complex because that temperature of reference involves only freshwater. In fact, because it contains salt, seawater freezes only below −1.8°C. And in certain circumstances, even very pure water does not easily become solid. This is the case with very small cloud droplets: they form from the condensation of water vapor and can remain in liquid form. Up to temperatures of −20°C, or even below, we then speak of supercooling. We should add that

Table 1.1.

Surface, volume, and contribution to the sea level of the different components of the cryosphere.

Component	Surface (millions of km^2)	Volume (millions of km^3)	Contribution to sea level (in m)*
Continental snow	1.9–45.2	0.0005–0.005	0.001–0.01
Sea ice	19–27	0.019–0.025	~ 0
Glaciers and ice caps	0.51–0.54	0.05–0.13	0.15–0.37
Ice shelves	1.5	0.7	~ 0
Ice sheets	14.0	27.6	63.9
Greenland	1.7	2.9	7.3
Antarctica	12.3	24.7	56.6
Frozen soils in winter	5.9–48.1	0.006–0.065	~ 0
Permafrost	22.8	0.011–0.037	0.03–0.10

Source: IPCC, *Climate Change 2007: Fourth Assessment Report* (Cambridge: Cambridge University Press, 2007).

Note: In the case of snow, frozen ground in winter, and the permafrost, the minimum and maximum levels are given. For glaciers and glacial ice caps, the two levels correspond to high and low estimates.

* The melting of one million km^3 of continental ice corresponds to a rise in sea level of 2.46 meters; the figures shown in this column take into account the replacement by seawater of the released surfaces below sea level.

water vapor exists in the atmosphere regardless of the temperature; it can condense on microdroplets that form around condensation nuclei, and it also contributes to the growth of snow crystals by passing directly from the vapor phase to the solid phase.

There are different types of precipitation that reach the ground in solid form. Two are anecdotal within the framework of this book because, except in unusual cases, they melt quickly. Graupels (much like a fine hail) come from the freezing of raindrops that travel through air that is colder nearer the ground. Hail and hailstones are formed higher up in the atmosphere, within cumulonimbus clouds, which reach an altitude of over 10 kilometers. Snow is formed both through the freezing of small supercooled drops and by direct condensation of vapor on preexisting crystals. Unlike graupels and hail, a large part of snowfall remains in its frozen state as long as the temperature of the surface on which it falls is below 0°C.

The thickness and extent of the snow cover vary depending on the season. Permanent snow is, of course, found in the mountains, on glacial ice caps, and on sea ice. Outside those zones, snow covers on average 26 million km², mostly on the continents of the Northern Hemisphere. At the end of January, that surface may reach 45 million km² but reduces to less than 2 million km² around the end of August. In the Southern Hemisphere, owing to the unfavorable distribution of the continents, snow fields cover scarcely more than .5 million km². This snow cover is fairly thin, a fraction of a meter on average. It thus represents a very small volume of ice.

This is the snow that, year after year, over centuries and millennia, feeds our white planet, which we call the cryosphere. On the continents, mountain glaciers, glacial ice caps (a term used when the ice surface is less than 50,000 km²), and the huge ice sheets of Greenland and Antarctica are the main elements of the cryosphere. We must also include frozen ground, which is permanent in some places (permafrost), and other frozen grounds that are frozen only in the winter, such as rivers and lakes. There is also sea ice, which is formed by the freezing of ocean water surfaces; icebergs that float on the edges of ice caps and ice sheets; and ice shelves, fed in some peripheral regions of Antarctica by the ice that flows from the interior of the continent.

Mountain Glaciers and Ice Caps

Outside the high latitudes, it is above all the winter snow and the mountain glaciers that symbolize the white planet. In most regions, when it rains on the plains it snows in high altitudes. The height at which snow is permanent varies between about 3,000 and 4,500 meters depending on the latitude and the site's orientation to the Sun. The density of the fresh snow cover is very low (0.1 to 0.2) because it contains a lot of air, which occupies the spaces left free between the complex lacy shapes of the snowflakes. The laws of thermodynamics dictate that these flakes will lose their edges through sublimation and assume a more spherical shape. This facilitates the piling of the layers whose density increases (more than 0.5 for the *névé* [snow that is more than a year old]). With recrystallization the smallest flakes disappear, giving way to the larger ones, and the amount of air captured between the crystals decreases. The ice thus formed has a density close to 0.9, the same as that made in a freezer. Both the névé and the ice then float on water. The amount of time required for these processes can vary considerably; it depends, for example, on the climatic conditions, the temperature, and precipitation. In the mountains, when the winter snow doesn't completely disappear during the following summer, it becomes névé. The percolation of the melted water can lead to the formation of ice in a few years' time.

There are approximately 160,000 such glaciers of all sizes. They are found on every continent except Australia. They cover some 430,000 km², and the largest are found in Alaska, in the Canadian Arctic, in the South American Andes, and in the Himalayas. On the other end of the scale, some African glaciers measure only 15 km². Among the longest glaciers are those of the Bagley Ice Field in Alaska (185 km), the Siachen in Karakorum (75 km), the Inybtehek in Tian Chan (65 km), the Uppsala in Patagonia (60 km), the Monaco in Spitzberg (60 km), the Tasman in New Zealand (28 km), and the Ngojumba in Nepal (22 km). Alpine glaciers cover around 3,000 km². Half of them are in Switzerland, including the Aletsch Glacier, which measures 25 km in length; 10% of this surface is in France, including the Mer de Glace (Ice Sea) near Chamonix, which flows over 12 km. Contrary to what the name "mountain glaciers" might suggest, some, depending on the climatic characteristics of their location (precipitation, temperature), can flow as far as the sea: this is the case of the Malaspina (Alaska) and the Darwin (Chile) glaciers. They can become as much as several hundred meters thick.

We generally use the term *glacial ice caps* for glaciers whose surface can vary from a few thousand to 50,000 km². They are found in high altitudes in polar and subpolar regions and are relatively flat. In Svalbard (Spitzberg), the largest ice cap found in the North-East Land extends over 8,000 km² and is 300–400 meters thick. In Iceland, two ice caps (each 1,000 km²) cover the summit of the island, and the Vatnajokull (8,300 km²) is made up of 3,500 km³ of ice. Other ice caps are found in the Arctic, in the northern Canadian archipelago, in Siberia, and in southern Chile. There are approximately seventy of various sizes.

Glaciers and ice caps cover an area that is close to that of France; their volume amounts to between 50,000 and 130,000 km³. The area of the glaciers and ice caps located on the edges of Antarctica and Greenland, not taken into account in Table 1.1, represents 140,000 km².

Polar Regions: The Omnipresence of the White Planet

Zero degree Celsius (0°C) in the middle of summer on the sea ice of the North Pole. Only nine days a year without frost in Resolute Bay (74° 43′ N, Canada) at sea level. A yearly average of −30°C at the center of Greenland at 3,000 meters in altitude. These figures provide the image of a region, the Arctic, where cold is the rule. On the other side of the planet, −11°C is the average temperature at the coastal base of Dumont of Urville (66° 07′ S) and −55.4°C in Vostok in the heart of Antarctica at 3,488 meters in altitude. And the temperature can plunge as low as −89°C in the winter. The polar regions hold unenviable records. The Sun has little love for them, and the feeling is mutual.

It is primarily the tilt of the Earth's axis of rotation (23° 27′) compared to the plane of its trajectory around the Sun that is responsible for the fact that the Sun has little love for the polar regions and keeps them among the coldest places on the planet. This tilt, also called the obliquity, determines the length of the presence of the Sun as a function of the latitude and thus the rhythm of the seasons. In the summer, boreal or austral, above 66° 33′ in the north or the south, the Sun can be continuously visible, whereas at the opposite pole it is below the horizon for the six months of winter. In addition, in the polar regions the Sun's beams never shine perpendicularly. For any given surface, the energy absorbed is that much more reduced. Thus the high latitudes, in terms of solar energy received, are at a disadvantage. As if annoyed by the lack of attention, the white planet that spreads out over the poles has chosen to ignore the Sun since it sends back up to 90% of its incidental rays over a zone covered with fresh snow. This albedo effect varies with the nature of the surface; it can go as low as 50% for old snow and, though it can be around 80% over a large part of the ice sheet, it varies over the sea ice from 80 to 50%, depending on the condition of the surface and how reflective it is. It is thus essential to wear sunglasses in these zones to avoid blinding because the eye knows the difference between an albedo of 80% and that of rocky surfaces (15 to 30%) or of ocean surfaces (5 to 15%). In terms of climate, it is the size of this albedo that is one of the main characteristics of snow and ice. Greatly reducing the absorption of solar rays that are already weak in the high latitudes, it is largely responsible for the extreme cold of these icy regions, both in oceanic and continental regions.

But the Sun and the albedo are not the only players. In high latitudes during the cold seasons the ocean is covered with a lid of ice. Without it, the Arctic Ocean, in the middle of which we find the North Pole (Figure 1.1), and the Southern Ocean, surrounding the Antarctic continent, transfer a large quantity of heat into the atmosphere, maintaining the temperature of the air close to that of the seawater. But when the ice cover is in place, the "ocean radiator" stops and the winter temperatures above the sea ice can reach −30°C or even lower. In the Arctic Ocean and the adjacent seas to the north, as well as in the Southern Ocean, it is common that, following summer, the temperature on the surface of polar waters is lower than −1.8°C, which means that the surface is cold enough to freeze into sea ice. The first ice crystals give birth to "pancake ice," which, in time, forms sea ice. This process requires rather long periods of freezing because the surface water, when it is cooled, becomes denser; it sinks and is replaced by warmer water. After reaching a certain thickness, the ice isolates the ocean from the atmosphere and slowly builds up further by freezing of the colder seawater underneath the ice. Waves and swells slow down this formation of sea ice, which can become one to two meters thick in the first year. Over the years, and taking into account summer melting, it can then increase up to three to four meters.

Summer brings melting and partial dislocation but in very different ways in the two hemispheres. In the south, the sea ice attached to the continent is exposed to the whims of the Southern Ocean; it extends up to 17 to 20 million km² in September—that is, to a surface equivalent to that of the entire Antarctic continent—and five times less in February, or 3 to 4 million km². In the Arctic, the sea ice is surrounded by continental surfaces that stabilize it; it extends over 14 to 16 million km² in March with ice that is a bit thicker than that around Antarctica. And, as we will see, it becomes three to four times smaller in September.

On continents with particularly harsh climates without glaciers, frozen ground, or permafrost, forms. When the average temperature goes below 0°C, ice is present at great depths. Mammoths and bison, dating from the last

Figure 1.1. The Arctic regions, indicating the deep ice core drilling sites in Greenland.

ice age, have been found in these natural freezers in Siberia and Alaska. Depending on the season, this ground can thaw on the surface, but the thickness of the permafrost can reach as deep as six hundred meters. The surface areas are also considerable in these regions: one quarter of the exposed land in the Northern Hemisphere including Alaska, Canada, and Russia but also certain regions in the high mountains are covered by permafrost. However, the zones with a great deal of ice scarcely cover 2 million km². These permafrost zones can extend under seas and oceans, as well as under glacial ice caps and ice sheets.

One need only fly over Siberia in the winter to grasp the importance that snow has in the everyday lives of Russians. It's not surprising that their language is particularly rich in terms describing snow in all its forms, nor is it surprising that they have bequeathed many expressions to us, including the word *zastrougi* to describe the very typical relief of the great expanses of snow as seen on the high plateaus of the ice sheets swept by winds. *Zastroug* is in fact an old Russian word that means "hand plane" (referring to the manual planing of wood), and *sastrugi* are the shapes on the surface of the snow that form through erosion and in the direction of the wind. Such shapes are similar to those that are seen on a plank that has been manually planed.

Greenland, Antarctica, and Ice Shelves

The largest and most voluminous glaciers are indisputably the *inlandsis* ("ice in the middle of land") as they are called in Danish. They are on a completely different scale: one is in the north (Greenland) (Figure 1.1) and the other in the south (Antarctica) (Figure 1.2), home of the South Pole. These two enormous ice masses, which cover around 11% of the total surface of all continents, weigh heavily upon the Earth's crust. They cover that crust and their weight pushes down the rock base by around 30%. Whereas Antarctica is more or less centered on the South Pole, Greenland is an island around 2,500 kilometers in length and located over latitudes of 60° and 80° N. Both have true rivers of ice that flow toward the sea in the form of emissary glaciers, or extend into the sea in the form of floating ice shelves, mainly in Antarctica. The profiles of the ice sheets are noticeably parabolic. On the relatively flat central parts, the wind sculpts little dunes that resemble those in dry deserts, the *zastrougis*. In the coastal zones, where the slope is greater and the ice currents marked, crevasses form that make it difficult to gain access to the inland ice sheets.

In Greenland, ice covers a surface of 1.76 million km², or 80% of the total surface of the island, which with 40,000 kilometers of coastland is the largest in the world. The ice sheet makes up the largest part (1.7 million km²), to which are added glaciers and glacial ice caps located on the

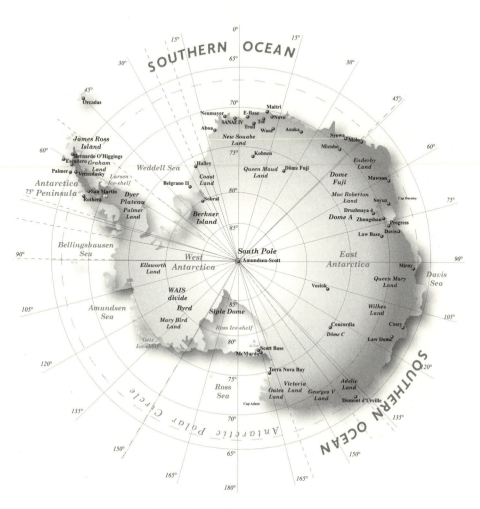

Figure 1.2. Antarctica, indicating deep ice core drilling sites.

periphery and on coastal islands. The surface of the ice sheet is characterized by the existence of two domes, one in the south reaching 2,873 meters and one in the north at 3,236 meters. The dividing lines of the ice
flowing to the coast depend on this asymmetry in altitude. There, toward the coast, myriad glaciers create paths through a discontinuous
and low mountain chain. The widest is the Humboldt Glacier located
in the northwest. The front is 120 meters high and 80 kilometers wide.

These glaciers isolate the *nunataks* ("land in the middle of glaciers"). The largest and the most active is the Jakobshavn Glacier on the west coast, around 70° N. It drains the ice from the ice sheet over a length of more than 300 kilometers. The icebergs it releases invade the fjord that the glacier has dug, which, even in the summer, assures the presence of sea ice. The icebergs that come from the largest ice rivers are of modest size and relatively regular shape. The fjords, which have been sculpted by the glaciers, are filled by icebergs over dozens of kilometers. The ice attains a thickness of more than 3,400 meters in the central regions, where the floor is a concave basin that seems to have been sunken under the weight of the ice. The ice becomes much thinner, or even nonexistent, when rocky outcrops appear in the coastal regions; it is then on average close to 1,800 meters thick. The volume of ice is around 2.9 million km^3.

While certain zones of the North Atlantic are relatively tempered by the Gulf Stream, this is not the case in Greenland, which has a continental climate. The average annual temperatures can be close to 0°C in the coastal regions, which causes the melting of ice in the summer. On the ice sheet, temperatures can become as low as −30°C in the central zones. One can also encounter surfaces where the snow contains greater or lesser amounts of water infiltrations that refreeze during the winter. The zone where the snow is "dry" is found in the central regions above 70° N. If, in general, temperatures decrease from the coast to the interior, snowfall is more abundant in the south where it can reach several meters. There is much less snowfall over the rest of the ice sheet where the annual layer decreases to about a dozen centimeters.

On the other side of the Earth, the surface of the Antarctic continent, 12.3 million km^2, or more than twenty times the area of France, is almost completely covered with ice. The exposed rocks on the Antarctic Peninsula, in the chain of Transantarctic Mountains and on some coastal ranges, represent only 1% of the continent. Geographically, we can describe the ice sheet as being made up of three parts. At 10 million km^2, East Antarctica is the largest; its central and most elevated part is relatively flat but rises to more than 4,000 meters. This ice rests on an ancient floor, a part of which is currently located above sea level but at below 1,000 meters in the Marie Byrd Land and reaching −2,540 meters in the Bentley Subglacial Trench. Given this very irregular floor, it is difficult to determine the thickness of the ice from the topography. The layer is much thinner in the coastal regions, but it reaches 4,800 meters at 40 kilometers away from the coast of Adélie Land, whereas the surface altitude is 2,400 meters. It is in East Antarctica that we find the true center of gravity of the ice sheet at around 81.5° S and 73° E; this point is the farthest away from the coasts and the conditions there are such that it has been dubbed "the pole of inaccessibility." The ice flows toward the sea and many glaciers find their path through the coastal mountain chains. The largest, the Lambert Glacier, is 120 kilometers wide and approximately 40 kilometers in length; it feeds the Amery Ice Shelf. But this ice also feeds West Antarctica through the trans-Antarctic chain, which culminates at Mount Kirkpatrick at an altitude above 4,500 meters. The topographies of the bedrock of these two areas are very different.

The area of West Antarctica (1.8 million km^2) is more complex since it includes several domes as high as 2,300 meters. The Antarctic Peninsula (around 0.5 million km^2, 300,000 of which are covered by ice) abuts West Antarctica and stretches away, thinning, toward South America. The surface elevation reaches up to 2,300 meters and the average thickness of the ice is on the order of 1,000 meters. In all, the continental ice in Antarctica represents a volume of 24.7 million km^3.

Antarctica is also characterized by the presence of many huge ice shelves fed by the ice flowing from the ice sheets. When the ice reaches the sea, it begins to float. The shelves on the perimeter of the Arctic Ocean are few in number and not very large. Favored by the topography of the subglacial floor that often descends below sea level at the edges of the glaciers, by the presence of wide bays fed by large glaciers, and by the existence of islands that serve as anchors, they occur in great numbers in Antarctica, where they occupy close to half the length of the coasts.

In West Antarctica two of them have surface areas comparable to that of France. The Ross Ice Shelf covers close to 500,000 km^2, with an average thickness of 430 meters. On the Antarctic Peninsula, the Ronne-Filchner Ice Shelf is a bit smaller (450,000 km^2), but the ice is somewhat thicker (660 meters). Ice shelves are also fed by the snowfall and, in certain zones, by the ice that is formed at the bottom by freezing seawater. The very flat surfaces are broken only near the anchoring points where crevasses mark the line where continental ice begins to float. The fronts of ice shelves are like cliffs from which icebergs sometimes break off. These bergs are called tabular icebergs, and their name indeed describes their shape. In all, the surface of the ice shelves covers a bit more than 1.5 million km².

The two large Ross and Ronne-Filchner ice shelves have an important hidden characteristic: they float. They also stabilize millions of cubic kilometers of ice in West Antarctica, most of which rests on the Earth's crust, well below sea level. This innate characteristic will perhaps not be so hidden in the future. We are indebted to a British geologist from Ohio State University in Columbus, John Mercer, for having been the first to point out the potential danger of collapse to the entire ice cap of West Antarctica by the rupture of the two floating shelves in the event of the warming of the planet following an increase in the greenhouse effect caused by human activities.[1] The consequences of this would be an accelerated melting of the ice sheet of West Antarctica and potentially a five-meter rise in sea level. Mercer's theory was met with skepticism in 1978, but this possibility has since been suggested by many other scientists.

Ice: An Agent and Indicator of Climate Change

The variations in the extent of snow- or ice-covered zones are amplifying the fluctuations in temperature on a global scale. A cooling leads to an extension of ice, whose strong albedo contributes to the maintaining of low temperatures; by contrast, a warming of the climate reduces the area of ice-covered surfaces, which increases zones with a weak albedo and thus contributes to an increase in temperatures. Ice cover, through its extent and thickness, is both an agent in the climatic system and an indicator of any change: first, on a seasonal scale when the expanses of snow cover or sea ice follow the changes of the thermometer; in the longer term, on the scale of decades, when the change involves the advance or the retreat of mountain glaciers and the expansion or reduction of their volume; and finally, on the level of centuries and millennia, with respect to the ice shelves, ice caps, and ice sheets of the polar regions.

There are many interactions between the cryosphere and the other components of the climatic system on a global scale. The cryosphere plays an important role in atmospheric and oceanic circulation. Simply stated, the cold air is denser and the low temperature zones are also those of high atmospheric pressure; similarly, deep oceanic circulation originates from the

high-latitude oceans. On a more regional scale, the snow cover or the presence of sea ice reduces the atmospheric exchanges with the continents and the oceans. Above the cold zones, the atmosphere contains much less water vapor, and it allows the infrared rays emitted by the Earth to pass through, back to space. The result, then, is all the more negative, as it leads to even lower temperatures.

Furthermore, the phase transition between water vapor and ice requires a large amount of energy. It takes close to 80 calories to melt a gram of ice that is already at 0°C. In the spring, as soon as the snow disappears, temperatures climb very quickly in calm weather; the calories previously used for melting snow then heat the atmosphere. Regarding the energy expended in the climatic system, a few figures speak for themselves: to heat the ocean by 1°C, it takes a thousand times more heat than to increase the temperature of the atmosphere by the same amount. It would take close to twice as much more to cause 28 million km^3 of ice, which currently constitutes our white planet, to melt.

The White Planet and Sea Levels

An agent within the climatic system, an indicator of its changes, the cryosphere also plays an important role in the variations in sea levels, one of the key variables for humanity in the evolution of the climate. The changes in the volume of continental ice directly influence sea levels. We will see that the warming that followed the end of the last ice age, some 20,000 years ago, caused an increase in sea levels of around 120 meters. More subtly, the melting of our mountain glaciers added to the thermal expansion of the oceans has contributed to an increase of 10 to 20 centimeters observed in the last century. In this realm, the behavior of the ice comprising the ice sheets represents one of the major uncertainties involving the future of our planet.

Sea ice and ice shelves that float on the seas do not affect the level of the water as they melt, just as melting ice cubes in a beverage do not raise the level of liquid in a glass. On the other hand, glaciers, ice caps, and ice sheets are formed from precipitation connected to the evaporation of ocean water. They represent a volume of ice close to 28 million km^3 (Table 1.1), or an equivalent amount of water (around 25 million km^3), ice having a density of close to 0.92. Spread over the ocean's surface, this volume represents a layer of around 70 meters. This doesn't mean that the sea would rise by as much, since a part, around 10%, of the ice of the ice sheets is currently below sea level and the new distribution of the masses would lead to a readjustment of the Earth's crust. Taking these readjustments into account, one can then estimate an increase of around 64 meters. Let us very clearly specify that the melting of all the existing ice is not predicted by any scenario of the evolution of our future climate. But before exploring further what the ice reveals about our climatic future, let's look at a bit of history.

CHAPTER 2

From Exploration to Scientific Observation

In the eighteenth century, the way in which an educated man perceived our white planet was quite different from that which has just been presented. Of course, people knew about the existence of mountain glaciers and eternal snows that covered the highest peaks, but no geographer imagined that the amount of snow could fluctuate over time. The Arctic—at least its peripheral regions—was not a completely virgin land since native peoples lived there, but no chronicle mentions anyone reaching the North Pole or traveling across all of Greenland. As for Antarctica, that continent was terra incognita. On January 17, 1773, James Cook became the first person to cross the polar circle, declaring upon his return from what was then his second expedition: "I went around the austral hemisphere, following a high latitude, and ran along it in order to irrefutably prove that no continent exists, unless it [the continent] is close to the pole and out of reach of the sailors."

Certain zones in the center of Antarctica remain largely unexplored to this day, but our knowledge of these polar regions has increased enormously in recent times. Their geography and topography keep very few secrets in this era of satellites, but progress has also reached more intimate aspects of the evolution of various components of the white planet: the flow and mass balance of mountain glaciers, polar ice caps, and ice sheets; the conditions prevailing at their base; the thickness and processes of formation of the ice shelf and permafrost; and so forth. We cannot resist the pleasure of including a few anecdotes about (and mentioning some names among those who became known through) the discovery then the exploration of these extreme regions. However, it is above all the scientific aspects of the explorations, the methods of observation used by researchers, and a few notable results that should be emphasized. We will leave aside for the moment one of the questions that quite naturally comes to mind: Does the recent evolution of the cryosphere independently suggest a climatic warming, as has been observed during the last few decades?

The Flow of Mountain Glaciers

In August 1820, three guides on their way to Mont Blanc were thrown into a crevasse by an avalanche. Their remains were found on the face of the Bossons Glacier, 3,500 meters lower, in August 1861. From this tragedy one can estimate an average speed of flow of the glacier of 180 meters per year. More recently, on September 19, 1991, a couple hiking in the Tyrol near the border of Italy and Austria found a human body at an altitude of 3,200 meters. The man was small (1.6 meters tall), his body was 5,000 years old, and the objects scattered around him, including an ax, were characteristic of the Bronze Age. Many questions arose: How was he able to be mummified quickly, which enabled his preservation? Why wasn't he crushed by the weight of the ice and dislocated by its currents? Even the slowest glaciers are renewed in a few centuries. In fact, the very great age of the man demonstrates that there are niches in the cavities of the rocky relief where the ice can stagnate for a long time.

Looking at mountain glaciers, we see huge white expanses near the peaks and lower down, a winding surface design, with crevasses and seracs, rocky moraines: this is what suggests a flowing, like calm, lazy rivers that are transformed into agitated currents. In the first half of the nineteenth century, the Swiss Louis Agassiz and then other observers intrigued by the glaciers measured their speed of displacement. They noted that the speeds are higher in the center than on the edges—and faster on the surface than on the bottom—because of the friction of the edges on the glacial valleys and as a result of contact with the rocky ground. Ice is not rigid matter; its plasticity causes it to change shape, through the effect of gravity, under the pressure of its own weight (1 m³ of ice weighs more than 900 kilograms) and by following the slope of the glacial valleys. The speed of flow varies along the glacier; it is higher when the slope is greater, in which case the glacier is thinner. Another important factor in the ice flow is the decrease in viscosity of the deep ice with the warming associated with the geothermal flux. With the melting that occurs at lower altitudes, water penetrates into the glacier, which, owing to the presence of a liquid layer at its base, can slide and sometimes cause catastrophic advances. The dynamics of a glacier are thus very dependent on the presence or the absence of melting water. This is what distinguishes temperate glaciers from cold glaciers. In practice, the latter exist only at very high altitudes and in high-latitude regions. In the Alps, a cold glacier at the summit becomes temperate as it flows toward its front.

Let's look at the Mer de Glace Glacier closely. It is the largest French glacier, which stretches up to 12 kilometers in length, between 3,900 and 1,400 meters in altitude, and covering an area of 40 km². The feeding basin extends into a cold zone, and the height of the snow can be considerable there, reaching eight meters at the end of winter in the upper part of the Vallée Blanche. The Glacier du Géant, by way of the Vallée Blanche, the Glacier de Leschaux, the Glacier de Talèfre, the Glacier des Périades issued from Mont Blanc du Tacul and from the high peak of the Géant—they all converge to form the Mer de Glace, some 15 kilometers wide. Overall, the zone of accumulation represents two-thirds of the surface. Lower, one enters into the zone of ablation marked by the disappearance of the névé and the presence of seracs due to changes in slope. With warmer temperatures, the ice layer can lose as much as a dozen meters in thickness per year, the terminal tongue of the glacier marking a front, the evolution of which can be followed visibly. The thickness of the glacier varies along its entire length from one to a few hundred meters (400 meters at the foot of Mont Blanc du Tacul) but much less in the terminal tongue. In this pathway, speeds vary with the slope and thickness of the glacier, which then assures continuity in the flow of the ice. Speeds measured over the last kilometers of the Mer de Glace indicate a rapid decrease from more than 200 to 100 meters per year upstream, where the slope is greatest, to 50 meters per year at the terminal tongue where the glacier becomes increasingly thin.

The position of this terminal tongue, the front of the glacier, fluctuates over time. And the length of this glacier, a relatively easy parameter to determine, was measured directly in many sites in the Alps beginning in the 1870s. This is not, however, a very good indicator of climatic variations. Bernard Francou and Christian Vincent emphasize this:[1] two neighboring glaciers located in the same massif and having an analogous climatic history can undergo very different fluctuations in length that are asynchronous. In fact, the front of a glacier evolves as a function of its size, the characteristics of the local topography, such as the slope and orientation to the Sun, and its flow conditions, in particular near the bedrock. From a glacier's variations in length, we cannot deduce whether the entire mass of ice is increasing or decreasing. Even if, outside periods of advance (1890–1900, 1910–30, and 1960–90), the retreat of Alpine glaciers is rather homogeneous, some do exactly as they please. However, if we look at rather long periods of time, that is, more than a decade, using a statistical approach over a given region, variations in length (Figure 2.1) provide a first rough indication—which must be used with caution—of the evolution of the climate. This approach, taken by Hans Oerlemans, reveals that a retreat of glaciers was beginning rather homogeneously on a global scale at the beginning of the nineteenth century, became almost universal beginning in 1850, and continued throughout the twentieth century.

Mass Balance: The Health of a Glacier

More than a glacier's advance or retreat, its health depends on its mass balance, that is, the difference between accumulation and ablation. At the end of summer the quantity of snow that has accumulated during the year is measured using markers or drillings a few meters deep, spread out on the surface: that quantity indicates any increase in the mass of the glacier. Markers and altimetric measurements in the zone of ablation thus enable an evaluation of any loss of mass. The oldest series is that of the Swiss Clariden Glacier, begun in 1914. However, such monitoring, which is very cumbersome, is restricted to only a few glaciers; in 2005, only a dozen throughout the world had been observed for more than forty-five years. A more sophisticated approach involves estimating the altitude of the entire surface of a glacier at a given moment using topographical maps drawn from aerial or land photography. This technique, called photogrammetry, enables the reconstitution of past variations in volume. For the French Alps, some photographs from the 1950s are useful in this regard. They are, however, much less precise than current aerial views that are of excellent quality. To go back further in time, to the beginning of the twentieth century in the Mont Blanc massif, old maps, drawn with extreme minutiae, can also be used. In the near future satellite images should be good enough to systematically determine the mass of glaciers.

This annual assessment depends on climatic variables, in this case on a regional scale, as seen in the similar tendencies observed in different glaciers on the same mountain massif. The response time, due to their inertia toward

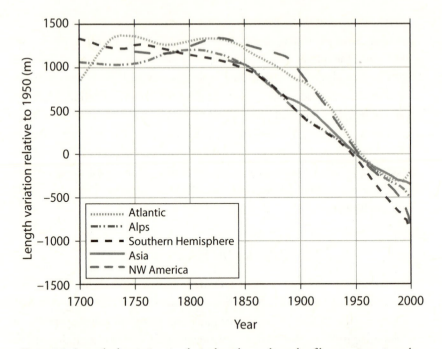

Figure 2.1. Smoothed variations in glacier lengths on the scale of large regions since the beginning of the eighteenth century. Glaciers are grouped as follows: Atlantic (South-Greenland, Iceland, Jan Mayen, Spitzberg, and Scandinavia), Alps, Southern Hemisphere (tropical glaciers, New Zealand, Patagonia), Asia (Caucasus and Central Asia), and North America (mainly the Canadian Rockies). Source: IPCC, *Climate Change 2007: Fourth Assessment Report* (Cambridge: Cambridge University Press, 2007)

climatic change, depends on the characteristics of the glaciers. By comparing the relationship between accumulation or ablation to the mass of the glacier, we arrive at values for the response times of about a dozen years for the smallest glaciers and of a century for the largest. More than the annual mass balance of glaciers, it is the cumulative mass balance since an initial date that is most often used to characterize a glacier. This cumulative mass balance is expressed in terms of variation of average thickness compared to the surface area of the glacier. Thus the Glacier de Saint-Sorlin has a negative mass balance; it has lost the equivalent of 44 meters of ice during the last 100 years. The mass balance is also negative for the other glaciers in the French Alps. This applies to all the Alps with a remarkably parallel trend in Austria, Switzerland, and France. In the Swiss Alps, it is estimated that the volume of the glaciers was reduced by nearly 50% between 1850 and 1999, from 107 to 55 km³, and that over the entire Alpine massif the surface areas of the glaciers have decreased by 40%, going from 3,800 to 2,500 km².

On the whole, the Alpine glaciers are retreating; this is also true, at least over the past fifty years, for other glaciers on the planet. The most marked shrinking has been seen in Patagonia and North America, then in Asia and in Arctic regions. This diminution is smaller and practically nonexistent in Europe, which is surprising after the picture we have just sketched of the Alps. In fact, the European assessment accounts for both the Alps and Scandinavia. The glaciers in northern Europe are distinguished by surprising new advances in the 1970s, a tendency that accelerated at the end of the 1980s. This advancing, limited to glaciers near the coasts, can be attributed to an increase in winter precipitation and in the amount of snow, all the more since a portion of the autumn rains have moved to the winter months.

Let's now look at the situation that threatens the tropical glaciers, those located in the northern part of the Andes, in East Africa, and in New Guinea. Looking at the Rwenzori Mountains, on the border of Congo and Uganda, the area of the glaciers there, now less than 1 km², is one-seventh what it was in 1906, and we expect them to disappear completely during the next two decades. The same diagnosis has been made by glaciologists regarding the mythical Kilimanjaro Glacier, but the causes of its marked shrinking are hotly debated: rather than a direct consequence of climate warming, the glacier's reduced size could be linked, at least up to a recent period, to a drier climate. The glaciers of the Andes are evolving similarly to those of East Africa: the cumulative losses over the recent period have been as much as one meter per year. Some small glaciers, like the Chacaltaya Glacier in Bolivia, have recently disappeared. The larger ones resist better, but for how long?

The Arctic Ocean in the Time of the Explorers

Beginning in the sixteenth century European rulers and merchants were looking for another route to China by going north. By boat and on land, passages were discovered, but the northwest routes, going around America, and the northeast ones, going along the coast of Siberia, were not easy to uncover in the middle of the ice. In his quest for a new route the explorer John Franklin set out from England in 1845 with 129 men and two ships, the *Erebus* and the *Terror*. Not one person returned alive. Their remains were discovered almost fifteen years later, skeletons of men walking in search of their salvation. It was the Norwegian Roald Amundsen, one of the great legends in Arctic exploration, who opened the way in 1905 after an expedition that lasted three years. After spending two winters in the Arctic, he was able to locate the surface magnetic pole and to study the Inuit way of life. The site of so many sacrifices, the Northwest Passage would remain unused for a long time.

The Finn Nils Nordenskjold opened the northeast route from Sweden to Japan in 1878–79, combining scientific agendas and economic motivations. But the use of the passages he discovered, which depends on the variable presence of ice, is difficult even today despite the shrinking sea ice.

Let's look again at the nineteenth century and the first explorations of the sea ice, which extends into the seas of the Arctic regions and, at least in the winter, occupies a large part of them. In 1879 the *New York Herald* wanted to offer its readers a report on the mysteries of the North Pole: Was it a sea or land? To answer that the newspaper sent the *Jeannette* to the north of the Bering Strait. It was crushed by the ice in June 1881 at 77° 15′ N. Three years later, Eskimos found fragments of the wreck on the southwest coast of Greenland. Carried by the sea ice, the debris had thus floated more than 5,000 kilometers over the entire Arctic Ocean, at an average speed of five kilometers per day. That journey inspired a young scientist.

On June 24, 1893, the Norwegian Fridtjof Nansen set off from the port of Bergen. He had had a boat built, the *Fram* ("Forward"), and he planned to explore the polar basin—not by navigating but by allowing the vessel to freeze into the ice and then floating with it to the North Pole. At the end of September he reached the edge of the pack ice at 77° 14′ N, beyond the mouth of the Lena. He thought he would get close to the pole, but the *Fram* drifted first toward the southeast, away from its objective. In December the direction of the drifting was reversed: the ship ended up at the same latitude as two months earlier. The crossing of the Arctic Ocean then began. One year later, the *Fram* had traveled approximately 50 kilometers closer to the North Pole, but it is probable that the trajectory did not go beyond 85° N. Nansen and his companion, Hjalmar Johansen, left the ship in March 1895 to reach the pole with dogs, sleds, and kayaks. They had to give up: hummocks—true sentinels of jagged ice caused by internal pressure to which the sea ice is subjected during its formation and movement—made the surface chaotic and extremely difficult to traverse. In the spring, the canals of flowing water increased, and the "moving carpet" of the sea ice did not go in the direction of the pole. On July 24, more than two years after their departure, they reached Franz Josef Land. They finally reached Spitzberg after spending another winter, surviving by killing bear and walrus like true Eskimos. On June 17, 1896, more than three years after the beginning of this exceptional adventure, everyone ended up, including the boat, at Tromsoe, in Norway. The *Fram*, which has sometimes been presented as the most resilient ship in the world, can still be seen today in the Oslo museum dedicated to it.

The Arctic Ocean: Vulnerable Ice

Since that time, satellite observations of markers placed on the sea ice and drifting stations occupied mainly by Russians and later by Americans have allowed scientists to describe the drifting of the Arctic ice pack. Two main currents control the ice. One is transpolar: it starts from the Bering Strait, goes along northern Russia, then goes toward northern Greenland and follows its east coast, bringing Arctic ice to the North Atlantic. The average speed of the ice is on the order of five kilometers per day; it is slower in the Arctic basin where the ice is thicker (two kilometers per day) but much faster (20 kilometers per day) in the Greenland Sea where the ice pack breaks apart. A branch of this current goes up toward northern Canada, and the Beaufort Sea feeds a circular current in the heart of the Arctic Ocean. This is where we find the oldest ice (more than ten years old) and the thickest (seven to eight meters thick). This thin layer, on an ocean thousands of meters deep, can appear quite vulnerable.

This is the case at least in the Arctic, where one of the sources of data are measurements taken by submarines equipped with sonar. Kept secret for a long time, they have recently been made available to the scientific community by the Russian and American militaries. The results of these observations are revealing: they indicate as much as a 40% decrease in the thickness of the ice. But since they are limited to the routes followed by the submarines, they do not provide a reliable estimate for the entire Arctic Ocean. However, combined with the results of models, the available data indicate a thinning that, since the end of the 1980s, might have reached one meter in the central regions of the Arctic. We expect a great deal of data to be made available in the near future from satellite methods based on radar or altimetry observations using lasers.

The fragility of the sea ice in the Arctic is also illustrated by the decrease in its surface area since 1978; we know this thanks to microwave measurements from satellites (Figure 2.2). This area has decreased notably—close to 3% every ten years—and even more quickly in the summer—more than 7% every ten years. The connection between this decrease—which accelerated abruptly in 2007, then was followed by an increase in 2008 and 2009, while maintaining levels that were greatly inferior to those at the end of the 1970s, but decreased again in 2011 to a value close to the minimal value of 2007—and climate warming requires investigation; we will return to this in part 3.

In the south, around Antarctica, the sea ice can be traversed only in the short summer season by ships transporting provisions to the bases there. Its area increases by a factor of five in August and September; between summer and winter, it goes from four to 20 million km². The sea ice can then extend more than 2,000 kilometers from the coast. The rapid breaking up of the ice in November and December is facilitated by the ocean current that circulates around Antarctica and disperses the ice that will melt farther to the north in warmer waters. The ice is generally less than 50 centimeters thick and, in the summer, a large part of the coast is free of ice. In certain zones that are little affected by the marine currents (notably near the Ross, Bellingshausen, Amundsen, and Weddell seas), the sea ice can survive for several years and reach a thickness of two meters. Unlike what is observed in the Arctic, the surface area around Antarctica has not varied significantly over the last few decades; as for its thickness, the data are insufficient to provide any estimates.

While the Arctic sea ice melts first on its surface, creating pools, that of the south melts more rapidly from its base and on its sides. From the action of the wind and the marine currents, the ice can break up and liberate vast expanses of water, like the *polynya* (a stretch of open water surrounded by sea ice) of 350,000 km² often present in the Weddell Sea and detected by satellite observations. It was in one of these zones in the Weddell Sea that the *Endurance*, the ship of Ernest Shackleton, became imprisoned by the ice pack on January 20, 1915. After drifting for ten months, the ship had to be abandoned; for five months the crew survived on a floe before setting out on three small boats, surviving encounters with southern storms and finally

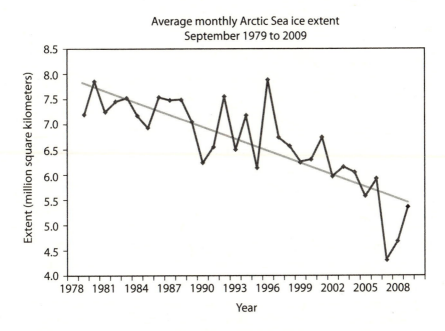

Figure 2.2. Variation in millions of square kilometers in the minimum surface area of sea ice in the Arctic Ocean between 1979 and 2009 with (straight line) the linear tendency. In 2011 the average monthly Arctic sea ice extent decreased again to a value close to the minimal value of 2007. Source: National Snow and Ice Data Center, Boulder, Colorado.

reaching Elephant Island in southern Georgia. This epic journey concluded with the rescue of the entire crew in August 1916 by a Chilean ship.

The evolution of sea ice is very dependent on the currents that circulate in shallow ocean waters. In the past few years, scientists have discovered another aspect of the interaction between ice and oceans: the formation and the presence of sea ice play a major role in ocean circulation and in the evolution of the climate. Cooling and the fact that the ice rejects a large portion of marine salt during its formation lead to denser surface waters, which sink to the bottom. They are replaced by warmer water coming from lower latitudes. This is a major feature of global ocean circulation through two zones of formation of deep water—in the north in the Norwegian Sea and in the south around the Weddell Sea. These zones thus play a key role in the "conveyer belt" that describes ocean circulation and whose behavior and intensity are important for the evolution of the climate, including that of the continents.

Greenland: An Island Inhabited for Millennia

Unlike Antarctica, Greenland has been inhabited for thousands of years. The first people to set foot on this huge, frozen, snow-covered island arrived more than 10,000 years ago from the North American continent, taking advantage of the low level of the sea. The harsh, cold climate did not prevent successive migratory flows, which led to the development of several civilizations during the last 4,500 years. The best known and the most extended was that of Thule (1000–1500 AD). Can we attribute the beginning of the Scandinavian colonization of Greenland in the tenth century to the violent character of Eric the Red? It is quite probable. Banished first from his native land of Norway then from Iceland, he continued his journey, traveling along the eastern coast of Greenland, and landed in the south, the more inhabitable part of this island. Was there no snow or ice visible to the Vikings at that time? Or did the island seem at least more favorable than Iceland, which appeared to them as an island of ice? Perhaps, like an early tourist agent, he was trying to attract Vikings to colonize this new land. Whatever the case may have been, he managed to persuade hundreds of people to follow him. Norway annexed Greenland in the thirteenth century and Denmark in the seventeenth century. Inuit populations and the Scandinavian colonists remained concentrated in villages confined to the coasts; as for the exploration of the inland ice sheet, that did not begin until the twentieth century.

Greenland: An Increasingly Negative Mass Balance

In 1912, the Swiss geophysicist Alfred de Quervain took the first glaciological measurements in Greenland when he traveled across it on skis. In 1932, two German scientists spent the winter on the ice at the center of Greenland at Eismitte Station at more than 3,000 meters in altitude: Alfred Wegener, father of the theory of continental drifting, who did not return from his journey, and Fritz Loewe, who lost a leg to gangrene. On the agenda: meteorology and glaciology in very difficult living conditions. Later the French polar expeditions of Paul-Émile Victor took the first measurements of the altitude and thickness of the inland ice sheet. From 1959 to 1974 the international glaciological expedition to Greenland led jointly by five European countries (including Denmark, the statutory godfather of the island) studied the mass balance of the ice cap. The Americans, within the framework of the International Geophysical Year but also motivated by strategic interests, were present on the field and developed, among other things, deep core drilling techniques in the ice;[2] this was the starting point for the study of glacial archives. Geodesic chaining on the ground enabled scientists to establish profiles of the surface and to obtain surface ice velocities. During fieldwork, the thickness of the ice was measured using gravimetry or seismic prospection, but, as with the Alpine glaciers, one of the most important objectives was to establish the mass balance and thus to map accumulation and ablation conditions.

Accumulation depends on the quantity of snow that falls. The quantity decreases as one moves to the north, reaching a maximum of 2.5 meters of equivalent water per year in the extreme southeast and less than 0.15 meters toward the northeastern end. It also diminishes from the coast toward the interior, as the altitude increases and the temperatures become colder. Over all of Greenland, the accumulation is estimated at 520 billion tons—520 gigatons per year (gt/yr)—or 30 centimeters of water on average. Nevertheless, these figures are only 5–10% accurate.

Ablation occurs with the melting of the snow, the water refreezing in the deeper cold layers or flowing in the form of rivers in the marginal zones, and with the calving of icebergs in the coastal regions. The equilibrium line separates the zones of accumulation and ablation at an altitude that varies from 1,000 meters in the north to 1,800 meters in the south. Unlike glaciers in the central regions, which have flatter topography, some coastal glaciers have very high flow speeds, as is suggested by the existence of fields of crevasses and fjords filled with icebergs. These glaciers drain from vast zones of accumulation. All along their path, one can estimate speeds assuming that the inland ice sheet is in a state of equilibrium. At each point along the length of a flowline, the quantity of ice transported must be equivalent to the accumulation upstream, minus the ablation in the coastal zones. Thus we calculate an average speed taking into account the thickness of the ice. Less than one meter per year in the central domes, these speeds increase progressively and reach up to several hundred meters per year in the terminal tongues flowing into the sea. In the coastal zones, the contrast is great between these rivers of ice that flow next to other ice, which is seemingly immobile, attached to the bedrock. These high speeds reveal a sliding due to the presence of water at the base of the glacier. This is hardly surprising in the zones of ablation, but this phenomenon is also encountered in certain zones of accumulation; they are colder, but the geothermal flux and the thickness of the ice can lead to higher temperatures in the ice that is in contact with the rocky base. The ice velocities calculated for a state of equilibrium are close to reality except in coastal zones and in those where there is sliding. They suggest that the dynamic response of the great basins of accumulation of the inland ice is typically of several millennia. Using an oversimplified calculation, we can compare the average accumulation over the entire inland ice sheet—0.3 meters per year—to the average thickness—1,800 meters of ice equaling 1,620 meters of water—leading us to arrive at an "average time of residence" of 5,400 years.

We estimate, with accuracy on the order of 10%, that the loss of mass in the form of rivers (streaming) is around 300 gt/yr and that the output of the glaciers in the form of icebergs is on the order of 235 gt/yr, with an accuracy of ± 15%; thus, the Jakobshavn Glacier alone provides 23 gt/yr, a value comparable to the output of our great French rivers. The accumulation of snow thus seems more or less balanced by the flow of meltwater and the formation of icebergs, which then melt in the sea. However, the uncertainty in the various components of the calculations shows that this assessment of the mass balance remains approximate such that this method of comparison between accumulation and ablation does not allow scientists to determine whether the volume of Greenland ice is increasing, is in balance, or is diminishing.

Measuring the variations in altitude over the entire ice sheet as precisely as possible and thereby deducing the variation in its volume is a way of getting around that difficulty. That, which was only a dream some dozen years ago, is in the process of becoming reality, thanks to the satellite observations that even enable us to "weigh" the ice mass that covers Greenland.

Everything began accidentally. Not being able to see the surface, helicopters and light planes crashed on the ground in Antarctica quite simply because, in certain frequencies, the radar altimeters penetrate the ice and are reflected on the bottom rather than the surface. This in fact enabled us to obtain a detailed topography of the surface and the bedrock of the ice sheets. In the 1960s the Tiros satellite, put into orbit by NASA and equipped with optic receivers, enabled us to map the surfaces covered with ice—on continents as well as on the oceans. The European Remote Sensing (ERS) satellite, one of the first designed for polar regions launched by the European Space Agency (ESA), is equipped not only with an altimeter but with a radar imager that measures the surface velocities of the ice through interferometry or by following identifiable markers. The positioning of markers on the ground by the GPS system also allows us to measure ice speeds at the surface. Over time, the radar data from the satellite altimeter can determine the evolution of the volume of glacial masses, since we know the altitude to within a meter of every two-kilometer span. The data from laser altimetry obtained from airborne measurements are also used, such as those taken in 1993–94 and then in 1998–99; however, these covered only certain regions of Greenland. Since 2002, the satellite instrument GRACE (Gravity Recovery and Climate Experiment) has provided measurements of the field of gravity and its variations over time. After many corrections, such as incorporating data connected to tides, the data acquired above Greenland have enabled us to evaluate the geographic distribution of ice and its variations over time.

GRACE has already provided spectacular results. These satellite data show that Greenland likely began to shrink in volume in the 1990s, a phenomenon that has recently accelerated. Between 2000 and 2008 it lost 1.5 billion tons of ice, contributing on average each year to a sea-level increase of 0.46 mm.[3] This contribution has increased during the past few years with an average level of 0.75 millimeters per year since 2006.

Antarctica: A Much More Recent Exploration

As a result of the intuition of thinkers, the maps of the Greeks pushed back far to the south the other summit of the world, that is, the counterpart of the Arctic. It was the Englishman James Cook who, on January 17, 1773, crossed the polar circle at 66° 33′ S and reported that there was no continent within reach of navigators. Seal and whale hunters had already ventured south of Cape Horn but kept their hunting grounds secret. It was Faddey Bellingshausen, on a mission for Tzar Alexander I, who earned the honor of discovering the continent in February 1820, while the sea ice was at its smallest. Twenty years later, three expeditions explored vast sections of the eastern coast. Among them, Jules Dumont d'Urville baptized Adélie Land with his wife's first name. His mission was to locate the magnetic South Pole. This was done by James Ross, who discovered Victoria Land a bit farther east and reached

77° S, at the foot of the ice shelf that bears his name. The extreme south was not mentioned for another fifty years, until it was stated during the London congress in 1895 that "the exploration of Antarctic regions is the most important geographic work to be undertaken before the end of the century."

The Belgian Adrien de Gerlache spent the winter of 1898 on his three-master caught in the sea ice, while the Norwegian Carstens Borchgrevink installed the first base on the continent. Other Europeans were there; the German ship *Gauss*, imprisoned in the ice, narrowly escaped the fate of the *Antarctic*, a Swedish ship crushed by the ice.

Jean-Baptiste Charcot spent the winter of 1904 onboard the *Français* and explored the coasts of the Antarctic Peninsula. He set off again in 1908, this time with the backing of the government, onboard the famous *Pourquoi Pas*. He spent the winter on a little island and explored the coasts, the sea ice, and the fauna during extensive fieldwork. He was one of the first to give priority to scientific objectives in an expedition. Charcot did not return to the south; he journeyed in the Arctic seas until his death with the disappearance of the *Pourquoi Pas* in 1936 during a storm off the coast of Iceland.

In December 1911 Amundsen was the first to reach the South Pole, just in front of the crew of the British explorer Robert Scott, whose five members never returned to the base camp. Shackleton got close to the pole in 1909 but backed off in time, saying in a message to his wife: "I thought that you would prefer a living ass to a dead lion." Shackleton did not give up, however, and soon set off on a new adventure: crossing Antarctica from the Weddell Sea to the Ross Sea, passing by the South Pole. Having left in August 1914, he didn't reach the continent because his ship, the *Endurance*, was caught in the ice in December. This was the beginning of an extraordinary voyage during which he was forced to abandon his boat, drifted on the sea ice, crossed wild oceans in lifeboats, and crossed a mountain chain in South Georgia. He returned to England at the end of 1916 without losing a single man. Shackleton died in 1922, once again en route to Antarctica. It is not surprising that he became a legend in polar adventure circles.

After the heroic period of discovery and of the first explorers came the era of observations and research, initiated by Charcot. Tractors, planes, airborne cartography, and telecommunications started to appear. In 1946 the American operation Highjump, directed by Admiral Richard Byrd, took

thousands of airborne photographs. At the end of 1949, the French polar expeditions of Paul-Émile Victor established the Port Martin base in Adélie Land, an observatory that functioned for three years until it was destroyed by fire. During that period, the Australians established a base on the Wilkes Coast, while the Norwegians, Swedish, and British spent the winter on the other side of the continent, not far from the Argentines and Chileans who were present on the Antarctic Peninsula. All these bases were established in the coastal zones and, with its 14 million km², Antarctica remained at this time a desert, especially since research was generally carried out in reaction to local events, such as the presence of a herd of emperor penguins or the blizzard on Adélie Land. Everything would change with the International Geophysical Year (1957–1958).

Twelve countries set up forty-eight stations, four of which, for the first time, were on the inland ice sheet: the Americans were at the South Pole, the Soviets at Vostok (the coldest place on Earth). The English on Queen Maud Land and the French on Adélie Land spent the winter in inland bases supplied by their coastal bases. The French, in the Dumont d'Urville and Charcot bases and during traverses, collected the first data on Adélie Land: they put markers in place and measured accumulations, thickness, and velocities of ice in the coastal zones. The glaciological fieldwork extended to the inland ice sheet: the English crossed the continent on tractors, from the Weddell Sea to the Ross Sea; the Australians, French, Japanese, and primarily the Soviets explored East Antarctica.

Antarctica: A Long Uncertain Mass Balance

Even more so than is the case for Greenland, and in spite of years of observation using classic measurement and satellite data, much remains to be done to understand the evolution of the Antarctic ice sheet that is the size of a continent. The trans-Antarctic chain borders and separates East Antarctica from West Antarctica. Starting from Victoria Land, it disappears under the ice and reappears intermittently in the Pensacola and Shackleton mountains and on the eastern edge of the Filchner Shelf. The ice moves depending on the slope of the surface, and the contour lines indicate that the ice of East Antarctica flows in part into the western zone, feeding the glaciers that flow into the Ross Sea or feed the Ross Shelf. The Antarctic Peninsula has a more temperate climate and attracts tourist boats with its accessibility and biological wealth. From a central plateau the ice flows in the eastern part toward the Weddell Sea, feeding, for example, the Larsen Shelf, while in the west, the glaciers can flow directly into the Bellingshausen Sea. Fed by the snow and the glaciers of the central regions, but also by local precipitations, the ice shelves have distinct dynamics because they float on the sea. In coastal regions the temperatures are lower than they are in Greenland, so that there is no significant surface melting and ablation is reduced to the output of the glaciers. In analyzing the change in mass of the ice sheet, it is customary to consider the eastern and western parts and the shelves separately.

From the elevated central regions, the contour lines allow us to demarcate the different basins of feeding that drain the ice toward the emissary glaciers. In East Antarctica, a dozen basins, whose sources go back a thousand kilometers upstream from the coasts, are identifiable; on these basins the accumulation of snow is on the order of 620 gt/yr, an amount that is a bit higher than the output of the glaciers that occupy 13% of the length of the coasts. The Lambert Glacier, the largest in the world, covers a surface area of close to twice that of France. The irregularities in the surface, revealing the tensions between the ice river and the much less dynamic neighboring zones, appear around 400 kilometers from the coast. The speed at the front of the glacier, which is 50 kilometers wide, can exceed one kilometer per year. It feeds the Amery Shelf, and its fluctuations in mass seem approximately in balance.

All of East Antarctica appears to be in a state of equilibrium, insofar as the few measurements of velocity carried out by satellite are in agreement with the calculated speeds of equilibrium. The variations in thickness measured from space have not been measured long enough and are not precise enough to challenge this conclusion: they have been available only since 1993, and the orbits of the satellites leave a great void above 82° S. These data nevertheless suggest a slight gain in mass of around 20 gt/yr, which could be connected to a slight increase in accumulation.

The situation is not as simple in West Antarctica, which is partially fed by the ice coming from East Antarctica. Furthermore, this region, whose uneven shores are largely open to different seas, is what is called a "marine" ice sheet because it rests on a bedrock located for the most part below sea level; thus glaciologists believe it is less stable than the East Antarctica ice sheet. All these characteristics cause some disequilibrium in the dynamic evolution of

West Antarctica, which in the past few years has lost a great amount of mass. Consequently, Antarctica, on the whole, could begin to decrease in volume, as is suggested by the satellite data provided by GRACE. According to these data,[4] Antarctica lost 104 gt/yr between 2002 and 2006, and more than double that (246 gt/yr) between 2006 and 2009. This loss of mass, and thus the contribution to the rise in sea level, would be close to that observed for Greenland. As for Greenland, these recent results indicate an acceleration that could, as we point out in part 3, be connected to climate warming.

Let's continue on our historical journey by going back even further in time.

CHAPTER 3

Ice through the Ages

Today around 90% of the ice on land is found on the Antarctic continent around the South Pole. The second largest mass is the ice sheet of Greenland, near the North Pole. The rest of the land ice, as we have seen, is spread among the smaller ice caps of the Canadian or Siberian Arctic and in the form of mountain glaciers that remain only in high altitudes in tropical or equatorial regions. The polar regions thus constitute the preferred habitat of the planet's ice.

The idea that the situation could have been different in the past and that in certain periods more ice was present on the continents is rather recent. First expressed in the first half of the nineteenth century, this theory aroused considerable disbelief in the field, and it took thirty years for its proponents, including Louis Agassiz, to convince the community of geologists of the existence of past glaciations.

The Time of the Pioneers

If you go to Chamonix today, there is a good chance that someone will tell you the story of the Mer de Glace, which reached the valley at the end of the nineteenth century. Since then, the glacier's front has retreated by a kilometer and a half, leaving in its wake erratic piles of rocks that make up moraines. The discovery of the existence of glaciations was born out of the investigation of such moraines. Until the eighteenth century, geologists, called diluvionists, believed that these rocks had been transported great distances by the flood described in the Bible. It wasn't until the following century that this explanation was questioned and the notion of great glaciations occurring in the past was proposed. John Imbrie, professor of oceanography at Brown University, combined his talents as a scientific storyteller with those of his daughter Katherine, a journalist, to recount this superb scientific saga.[1] They showed how the idea of glaciations evolved at a time when the notion that colder periods had existed in the past was simply unimaginable.

The idea of past glaciations also germinated in the minds of several mountain-dwellers who, no doubt daily, observed the erratic blocks of stone, and it had already been suggested by several authors at the end of the eighteenth century and then at the beginning of the nineteenth. But Agassiz, in 1837, was the first scientist to defend the idea that these round rocky masses, found in places far from their point of origin, had not been transported by the flood but by glaciers that, at a time when the climate was much colder, had advanced far beyond their current boundaries. Many geologists remained skeptical. This is understandable because at the time no one knew of the existence of ice caps and ice sheets; it wasn't until 1852 that people learned that Greenland was a huge mountain of ice; it was even later for Antarctica.

Others, quickly convinced of the existence of such a glacial period, turned into true detectives. They traveled the world in search of random boulders. Over the course of approximately forty years they were able to pinpoint the regions that in the past had been covered with ice. And they were huge regions: a large part of the North American continent, the northern United States and Canada, part of northern Europe, the British Isles, Scandinavia, northern France, and part of Siberia. A study of the mountain massifs located within very defined perimeters enabled scientists to estimate the maximum thickness of the ice; some massifs had rounded shapes that had been completely covered; others, as is seen by the presence above a certain altitude of sharp and very irregular peaks, had been only partially covered. And the corresponding volumes of ice are impressive; on the continents of the Northern Hemisphere, there were, we now know rather precisely, as much as 50 million km^3 of ice, almost double the quantity that currently covers the Antarctic continent. Consequently, as evaluated correctly in 1868, the sea level was then some hundred meters lower than its current level. The geography of the coastal regions was thus rather different in certain regions from what we know today.

There were abundant theories. John and Katherine Imbrie discuss eight of them, developed between the middle of the nineteenth century and the 1960s. One of the earliest theories was based on the idea that this enormous quantity of ice was present because the Sun at that time shined much less intensely. Three of the theories, including that one, were immediately rejected, and the remaining five could not be tested, at least until recently. Among the latter there was the suggestion of a connection between glaciation and the concentration of carbon dioxide in the atmosphere which, we will later see, was advanced by the Swede Svante Arrhenius in 1896. It is, moreover, rather interesting to note that in 1986, the date of the publication of their book, John and Katherine Imbrie placed that theory in the category of ones that could not yet be tested. Thanks to the great ice core drillings in Antarctica, whose history we will look at in a later chapter, we now know much more about the connection between the climate and the greenhouse effect in the past.

It was only five years after Louis Agassiz had proposed the idea of a large advance of ice in the past that the theory (which was definitively established in the twentieth century) of a connection between the position of the Earth on its orbit around the Sun and this expanse of ice, henceforth known by the name of astronomic theory, was born. In 1842, the French scientist Joseph Alphonse Adhémar proposed this idea, then the Scottish scientist James Croll formalized it in a book that was published in 1875, and the Serbian mathematician Milutin Milankovitch gave it his seal of approval some fifty years later. We will soon look at the evolution of this theory whose validity was confirmed in 1976 from the analysis of marine sediment.

Let us emphasize that in Croll's time astronomic theory had been rather favorably received because it foresaw that in the rhythm of our planet's path around the Sun, there must have been not one glaciation but a succession of glacial periods interrupted by warmer periods. This is in fact what geologists had proven, noting the presence of different layers of matter, clearly of glacial origin, separated by soil in which one could identify pollen or vegetal matter undeniably appearing at times when the climate was milder. In 1909 the German geographers Albrecht Penck and Edouard Bruckner named the last four great glaciations that occurred in Europe after the names of German valleys: the Wurm, the Riss, the Mindel, and the Gunz. By examining the thickness of the sediments of the valleys, these two scientists estimated that the last glacial period occurred 20,000 years ago, which has proven to be correct. But at the same time that Croll was developing the bases of astronomic theory, other geologists were interested in the deeper rocks with the objective of describing the history of our planet as completely as possible. Between 1830 and 1865 the American Charles Lyell introduced a series of names to divide this history into eras and periods with lengths that were then unknown but were rightly thought to have been very long. Another mystery: glacial periods had taken place in the very distant past since traces of them were found in rocks from the Precambrian, the first period in the history of the Earth. The adventure of this ice of long ago, that present before the Quaternary, which covers the last two million years, is now relatively well-known but stills holds some surprises.

Ice of Long Ago

At least two ingredients are necessary for ice sheets to form on our planet: cold and enough precipitation in the form of snow in regions where there are continents. The temperature curve of the Earth's surface since its birth 4.6 billion years ago (Figure 3.1) becomes more precise as we get closer to the current era. It shows that the Earth's climate has been colder than it is today and thus conducive to a covering of ice greater than that of today—but only about 20% of the time. The appearance, evolution, and disappearance of glaciations throughout the history of our planet have been conditioned by the drifting of continents and by the evolution of the parameters that influence the energy available to heat our atmosphere to a greater or lesser degree: the luminosity of the Sun, the distance between the Sun and the Earth depending on the Earth's orbit, and changes that have occurred in the composition of the atmosphere (greenhouse gases, aerosols).

Since our Earth was born, the luminosity of the Sun has increased by 25–30%. The brilliant American astronomer Carl Sagan (1934–96) played an essential role in the development of the American space program and in popularizing science. In an article published in 1972 with George Mullen he stressed the paradox of the pale Sun. Its weak solar luminosity would have caused the Earth to resemble a ball of ice during its first two billion years of existence. However, the geological and paleontological evidence tells us that running water and living matter had already appeared on the surface of the planet. Other factors thus influenced the primitive climate, and we suspect that large quantities of greenhouse gas, water vapor tied to the presence of oceans, and carbon dioxide exhaled by the Earth's crust were present in the atmosphere and compensated for the solar deficit.

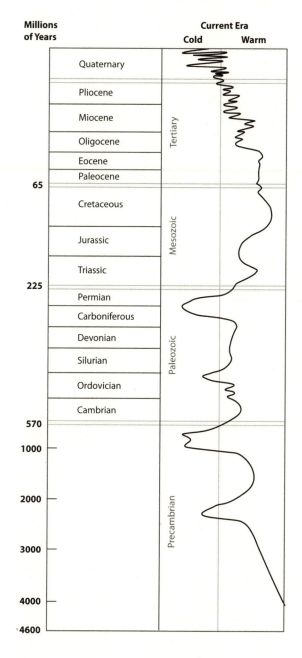

Figure 3.1. Temperatures throughout geological history. Source: Sylvie Joussaume, *Climat d'hier à demain* (Paris: CNRS Éditions/CEA, 2000).

As we mentioned earlier, the climatic history of our planet is essentially "warm." It was interrupted only from time to time by a small number of short (on a geological scale) periods characterized by more or less intense glaciations. The first to have been recorded dates from around 2.3 billion years ago, as proven by the stria that have been observed in rocks dating from that distant time and identified in many places. Indeed, only the friction of the ice could have sculpted such grooves on these rocks.

Then, it seems, there was nothing more of note; there were no known glaciations before 900 million years ago. Since then we encounter a past that was relatively rich with glaciations but also with very hot periods without ice. Thus there is strong evidence that the submerged land was substantially iced over several times between about 750 and 600 million years ago; our planet really did resemble a true snowball. At the dawn of that period, the land was but one supercontinent that was beginning to fragment. Almost all the continents were then gathered in the Southern Hemisphere, abandoning the Northern Hemisphere to a vast ocean. An a priori paradoxical situation, the ice sheets would have on at least two occasions reached the tropical or even equatorial regions.

How can we reach these conclusions for such a distant era? Early on geologists noted that many glacial deposits were preserved in layers formed in tropical latitudes during that era; the same was true for rocks exhibiting the characteristics of debris taken out of the ground by the friction of the moving glacier and then released on the bottom of the ocean by icebergs calved by the glacier. Paleomagneticians then came onto the scene tracing the alignment of the magnetic dust contained in glacial deposits. That alignment is governed by the Earth's magnetic field, which orients magnetic minerals in its direction. It proves to be an ingenious tracer of the latitude at which geological deposits were formed. Since the lines of this field go from one pole of the planet to the other, they align the magnetic material parallel to the surface—horizontally—at the equator and perpendicularly—vertically—at the poles. Thus the alignments of the glacial deposits of this era, the Neo-Proterozoic, suggest a formation at very low latitudes, tropical and equatorial. But it still had to be proven that in the course of the past 700 million years the magnetic signature of these rocks was not altered chemically or by any other process. This has been proven since the geological layers record several marked transitions in the latitude of the deposits, which would not have been observed in the case of a remagnetization of the material after its formation.

The paradox of a glaciation that extended as far as the equator motivated theoreticians and global climate modelers as early as in the 1960s. But the idea of "Snowball Earth," a term invented in 1992 by Joseph Kirschvink,[2] a geobiologist at the California Institute of Technology, took root out of observations in low latitudes suggesting not only that there was ice on the continents but that there were also frozen oceans. In fact, the almost permanent covering of the ocean by a layer of ice acted like a huge lid protecting the seawater from any exchange with the atmosphere, and the composition of the ocean was modified because of this; the sediments deposited on the ocean floor have retained the imprint of such modification.

These unusual periods remain marked by a certain mystery. Some scientists think that continents and oceans were covered with a true shield of ice; others believe there were free waters. How could there have been a world so different from the one we know today? And, above all, how did our planet manage to survive such change? The formation of the "snowball" would have been linked to the fact that the luminosity of the Sun, 6% weaker than today, was combined with low levels of carbon dioxide concentrations. Over these very long time periods, the erosion that brought carbonate sediments to the ocean represented a very effective carbon dioxide pump. According to Yannick Donnadieu[3] in Saclay and his colleagues, around 750 million years ago the relatively large parceling of the continent called Rodinia would have favored running water, thus erosion. The result: less of a greenhouse effect and a progression of the ice toward the equator that nothing could stop because the ice reflected an increasing amount of solar energy into space.

Greenhouse gases may have entered the scene again and helped the Earth escape from this "snowball" condition. With a layer of ice, there was no erosion of rocks and thus no trapped carbon dioxide being released into the atmosphere; however, volcanoes remained active and continued to emit carbon dioxide, which accumulated in the atmosphere to a point where it overpowered the reflective abilities of an ice-covered planet. Climatic simulations reveal that this required enormous concentrations of carbon dioxide, 350 times more than now, on the order of 10%. These figures are perplexing, and other theories implicate methane, another greenhouse gas. The hypothesis of a variation in the Earth's axis of incline was also advanced: strong angles of incline would favor colder temperatures in tropical and equatorial regions. This theory seems, however, to have lost its credibility. The debate goes on.

The luminosity of the Sun gradually approached its current levels, thereby eliminating in a certain way this risk of extraordinary glaciations. The climate was characterized by long, hot, humid periods interrupted by two glacial episodes, one at the end of the Ordovician, the other at the beginning of the Permian, around 450 and 300 million years ago, respectively. The ice sheets then developed in the polar regions, even if their geological traces are visible today at lower latitudes. Thus, at the end of the Ordovician, current West Africa was located at the South Pole in the middle of a huge continent, Gondwana, which encompassed Africa, South America, Antarctica, and Australia. So it is in the Sahara that traces of this Ordovician glaciation—extremely well preserved with an extensive ice sheet centered on the south of Hoggar—have been found, though (with a surface area estimated at 8 million km^2) it was less extensive than Antarctica's current ice sheet.

After the Permian glaciation, the climate became warm again—indeed, very warm, probably an average of 6°C above the current temperature—around 100–50 million years ago. Fifty-five million years ago there was an abrupt warming, the Paleocene-Eocene Thermal Maximum, which was likely connected to a large increase of either methane, due to the decomposition of clathrates, or carbon dioxide connected to intense volcanic activity. Then the climate gradually became colder; the cold reached Antarctica, which gradually moved farther away from Australia, taking position around the South Pole. This enabled the glaciers, and even isolated ice sheets, to settle at least temporarily on the continent. Of course we cannot observe the signs that they left on the bedrock because today it is covered with a sheath of ice that is several kilometers thick in places. But they calved icebergs into the neighboring seas—icebergs which, in melting, deposited grains of sand and rocks torn from the ground by the ice during its journey to the sea. Paleoceanographers indeed find debris dating from that era in marine sediment that they extract from the bottom of the Southern Ocean.

In the temperature curve in Figure 3.1 we immediately notice the particularly abrupt (on the scale of geological time) cooling that occurred around 34 million years ago. That was the transition between the Eocene and the Oligocene. This was the era when the first ice sheet covering almost the entire Antarctic continent appeared. The glaciation of Antarctica constitutes a major event in the climatic history of our planet, and it has become a favorite subject among paleoclimatologists. Was it the "abrupt" global cooling of the planet that gave birth to that huge inland ice sheet, or did the latter, once formed, change the climate?

Interest in this event goes back to the 1970s. In analyzing microfossils preserved in the deep ocean sediment, two paleoceanographers, Nick Shackleton and James Kennett, revealed the global cooling that had occurred in the course of the last 65 million years and identified the step that corresponded to the Eocene-Oligocene transition. Soon afterward, Kennett associated this event with the separation of Antarctica, Australia, and South America. Begun some tens of millions of years earlier, this separation enabled the formation of a circum-Antarctic marine current suddenly isolating the continent from the flow of heat coming from more tropical latitudes and thus putting into motion the cooling of Antarctica. A plausible explanation, but one that might not be the only one, as proven by climatic simulations published in 2003.[4] This work shows that the likely reduction of the atmospheric concentration of carbon dioxide by a factor of two in 10 million years would have had a noticeably greater impact on the cooling of Antarctica and the establishment of the ice sheet than that of the opening of the Drake Passage between South America and Antarctica. Thus the first large ice sheet of Antarctica was born perhaps under the preponderant influence of carbon dioxide, whereas the Southern Ocean was also being formed. By contrast, it was probably only 15 million years ago that a permanent ice sheet would have covered Antarctica with perhaps (here specialists are divided) extreme phases of deglaciation since then.

While Antarctica was separating from Australia and taking its position centered on the South Pole, the continents of North America and Eurasia were getting closer to the North Pole. More and more land was thus assembling in the high latitudes, both in the north and the south, and the ice of Antarctica reflected an increasing amount of energy into space. It thus participated in the cooling of the surface of the planet, likely helped by a decrease in the concentration of atmospheric carbon dioxide. Around 10 million years ago, the glaciers began to appear in several regions of northern high latitude, such as Alaska. But it was only in the Quaternary, the last three million years during which the Earth existed in its current topography, that true ice sheets developed in the northern regions of the Northern Hemisphere, including the Greenland ice sheet, the only large ice sheet that still exists in the north today.

Glaciations of the Quaternary and Astronomic Theory

The northern ice sheets then initiated a long series of fluctuations, whose rhythm was gradually modified with a dominant periodicity of nearly 40,000 years for up to around 1.2 million years. It then evolved toward a periodicity close to 100,000 years, with very pronounced glacial and interglacial periods for the last 500,000 years or so. The last million years have been the subject of a good deal of research, owing to an impressive wealth of marine sediment cores from the bottom of the oceans. They have enabled us to precisely define the rhythm of oscillations leading to the successive growth and disappearance of the large ice sheets of the Northern Hemisphere and to confirm the existence of a connection between the rhythm of the glaciations during the Quaternary and the position of the Earth on its orbit around the Sun.

Astronomers such as Johann Kepler, who in the seventeenth century demonstrated that this orbit is not a circle but a slightly flattened ellipse, are thus indispensable allies for anyone interested in the earthly climate or in that of the other planets. We must know everything about the trajectory of the Earth to be able to calculate the amount of solar radiation the Earth receives in a given place throughout an entire year (insolation), to understand that it varies from one year to the next, in a very slow but irremediable way, and to precisely determine these variations both past and future. This measure of insolation depends on three so-called astronomical parameters: eccentricity, obliquity (or tilt), and the precession of the equinoxes (Figure 3.2).

The eccentricity is quite simply the degree of flattening of the Earth's orbit. It varies from zero, when the orbit is circular, to around 6%, with periodicities close to 100,000 to 400,000 years. It is currently 1.7%, and the Earth is closer to the Sun in December than in July. The Earth's axis is more or less inclined in relation to the plane of the orbit, and that incline is called obliquity; its current value of 23° 27′ varies at a range of 2° 30′ following a cycle of 41,000 years. Precession is a more subtle parameter. It describes the fact that the place where our planet is located at a given moment of the year

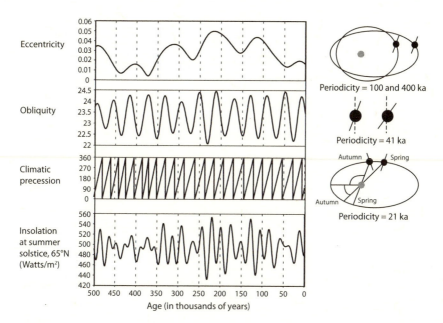

Figure 3.2. Variations in the Earth's orbit and astronomic parameters, eccentricity, obliquity, and precession. The lowest curve represents the insolation of the summer solstice at a latitude of 65°N.

moves over time: 11,000 years ago the Earth on June 22 passed by the point nearest the Sun, the perihelia; now this passage takes place on December 22. It will be on June 22 again in 11,000 years, which corresponds to a periodicity a bit greater than 20,000 years.

At any given moment, the insolation varies only as a function of latitude. On average, over long periods of time it depends only on eccentricity and varies very little—by less than 0.2% during the last million years. Its yearly variations are modulated by obliquity with opposite signs between the low and high latitudes, but these yearly variations remain relatively weak, not going beyond 6 Wm^{-2} (watts per square meter) at any latitude. In contrast, where there are large variations, they are on a seasonal scale: linked to precession, they can reach 20%. Let's take the example of Paris, which at the beginning of summer, 125,000 years ago, benefited from 458 Wm^{-2} but, 11,000 years later, had to be content with only 385 Wm^{-2}; today we are between the two with a value of 411 Wm^{-2}. Let us add that this seasonal variability is itself

influenced by the shape of the orbit: the variability increases the more ellipti-
cal the orbit.

Adhémar and Croll were correct in suggesting that the way in which the
Earth turns around the Sun helps explain the onset of the glacial periods. But
neither of them succeeded in establishing a theory that stood the test of time.
Adhémar postulated that the glacial periods were simply ruled by precession
and its close to 20,000-year periodicity and that they occurred in the hemisphere in which winter was the longest, returning alternately in the north
and the south. Croll believed only the distance between the Earth and the
Sun, and that value on December 21, the winter solstice, played a role; his
idea was that when that distance exceeded a certain value, the winters in the
Northern Hemisphere were sufficiently cold for a glacial period to begin.
Both Adhémar and Croll correlated glacial periods with very cold winters in
high latitudes. This was reasonable, it seems, but it does not stand up to analysis: temperatures are currently cold enough for Arctic regions to be snowy
in the winter without our being in a glacial period.

A decisive step toward determining the origin of glacial periods was taken
a few decades later by the Serbian mathematician Milutin Milankovitch,
who spent more than thirty years developing the theory that bears his name.
One of his objectives was to calculate variations in the insolation over long
periods of time for different latitudes, which is not a small task. Even though
he had the values of the three astronomical parameters over the last million
years, whereas in Croll's time they had only been calculated over 100,000
years, he had to devote almost ten years to his calculations, since at the time
he had just paper and pencils. In 1920, he achieved his goal and the results
were published; they attracted the attention of the German climatologist
Wladimir Köppen, famous for the classification of climatic zones from the
types of vegetation found in them. With his son-in-law, Alfred Wegener, the
father of the theory of continental drift, he was interested in the history of
climates and invited Milankovitch to Germany. The collaboration that resulted proved fruitful since Köppen provided Milankovitch with the key to
his theory: the triggering factor was a succession of cold summers in high
latitudes of the Northern Hemisphere, cold enough so that the snow of the
preceding winter did not melt. The snow, with its strong reflective power,
amplified the cooling to the point that a glaciation could commence. Milankovitch began to calculate the insolation in the summer at 55°, 60°, and 65°

N, and Köppen showed that the data had enough similarities with those of Penck and Bruckner (who described the succession of glacial periods in the Alps) such that Milankovitch continued to put the finishing touches on his theory and in 1941 published his defining book: *Canon of Insolation and the Ice Age Problem.*[5]

However seductive it may have been, the validity of this theory was not proven. There were no reliable chronologies, and there was no continuous record of past climatic conditions. These obstacles would be overcome thanks to the analysis of marine sediments that provided such a record over hundreds of thousands of years and could be correctly dated. In 1976, Jim Hayes, John Imbrie, and Nick Shackleton[6] provided this proof of a connection between the insolation and the rhythm of glaciations by demonstrating that the periodicities corresponding to precession and obliquity were present in the climatic series deduced from the analysis of marine sediments. Later the calculations of the amount of insolation were refined in relation to those of Milankovitch by André Berger in Belgium and Jacques Laskar in France. Berger showed that precession is in fact characterized by two periodicities, around 19,000 and 23,000 years, and that double periodicity was also present in the data.

Milankovitch earned well-deserved recognition, but his work was far from establishing the theory of paleoclimates. Many questions remained. And then, a few years later, the deep ice core drillings in Antarctica provided an additional dimension by revealing a very close connection between the composition of the atmosphere and the earthly climate over that time period. The drilling carried out in Greenland revealed extremely rapid variations in the climate that could occur on the scale of a few decades, or less, too rapid to be explained by the much slower variations in the Earth's orbit.

We will now turn to this fascinating scientific adventure: deciphering glacial archives.

PART TWO

POLAR ICE: AMAZING ARCHIVES

CHAPTER 4

Reconstructing the Climates of the Past

Human nature, it seems, has caused people to explore the story of our past ever since the dawn of time, whether it has been the history of nations, that of civilizations, or that of our planet and solar system. The history of past climates is no exception. This need of memory has become even more crucial over the last few decades, as an understanding of the climate and its evolution has become of major concern for the near future of our civilization.

To aid us in this task, we have meteorological measurements, which we've had for around a century. More recently we've been able to use observations from space. On the other hand, glacial archives, marine and lake sediments, fossil fauna and flora, soil, trees, limestone formations, and coral, have preserved a record of the past. When dated, these sources enable us to explore the past and to reconstruct continuously certain phenomena that, depending on the situation, describe the seasons, centuries, and millennia of the climate, of our atmosphere, of the oceans, and of the continents. The information thus archived requires a great deal of deciphering, in a certain sense analogous to the task of Jean-François Champollion in the nineteenth century when he unraveled Egyptian hieroglyphs. The climates of the past are the subject of a distinct scientific discipline: paleoclimatology.

The guiding principle of paleoclimatology rests on a simple idea: many characteristics of what exists on Earth, whether in the living world, in the mineral world, or in the water cycle, depend on the climatic conditions that existed when the different elements were formed. If these diverse materials accumulated layer by layer in such a way that the sequence of time of the deposits has been preserved, at least for the most part, we have then a climatic archive that can be used, provided that it is physically accessible, that it can be dated, and that some of its properties can grant us access to information that enriches our understanding of the climates of the past. The methods that are used to analyze these archives, adapted to the environment and the period under study, are extremely varied.

The simplest method, at least in principle, is studying the distribution of species, whether animal or vegetal, or for a given species, its annual growth pattern. We have here an obvious potential climatic indicator. Another method we can use, perhaps less concrete for laypeople, is the chemical composition of a matter. Whether it comes from the mineral world or that of the living, it is likely to be influenced by climatic parameters and is a very interesting source of information. We imagine that, for many readers, the term *isotope* is shrouded in mystery. And yet isotopic analysis is, by far, one of the most efficacious tools of the paleoclimatologist, whether he or she is studying continental, oceanic, or glacial archives.

The Round of Isotopes

We are all familiar with the terms *atom*, *hydrogen*, *oxygen*, *carbon*, and *nitrogen*, as well as the elements that make up Mendeleev's table. Each one is characterized by an atomic number that represents its number of protons and electrons: 1 for hydrogen, 6 for carbon, and so forth. A certain number of neutrons can be added to the number of protons (the atomic mass corresponds to the sum of protons and neutrons). A given atom can have various values: these are called isotopes. Let's take the case of hydrogen. It corresponds by more than 99.985% to an atom formed by a proton and an electron, but in around 0.015% of the cases a neutron is added to form an isotope of hydrogen that is commonly called deuterium, the one that creates so-called heavy water. Add another neutron and we have tritium, a radioactive isotope present in infinitesimal quantities in nature, whereas deuterium is a stable isotope. Carbon also has three natural isotopes: the main one is carbon 12 (with 6 neutrons), and the two others are carbon 13 and carbon 14; the latter is radioactive. Oxygen's three natural isotopes are oxygen 16 (8 neutrons), which is the principal atom, oxygen 18, and oxygen 17; the latter two are stable.

The chemical properties of the isotopes of the same element are identical, but their physical properties are slightly different. These minor differences put their stamp on all the mechanisms that have taken or take place in our universe whether during the formation of our planet or, nearer to us, when a cloud creates rain, hail, or snow. The radioactive isotopes are able to tell us the age of an event, to date it, and in some cases to describe its duration. In fact, a true discipline has been built around isotopes—isotopic geochemistry—whose applications are particularly prolific in the realm of paleoclimatology, thanks primarily to the isotopes of hydrogen, carbon, and oxygen that we have just mentioned.

Let's look at water. Most water is made up of $H_2{}^{16}O$ molecules, formed by two hydrogen atoms (of mass 1) and one oxygen atom of mass 16. The existence of deuterium and oxygen 18, the most abundant isotopes of hydrogen and oxygen, leads to "isotopic" forms of the water molecule, HDO and $H_2{}^{18}O$, which accompany $H_2{}^{16}O$ in all reservoirs where water is present in average concentrations close to 0.03% for HDO and of 0.2% for $H_2{}^{18}O$. When we analyze a sample of rain or snow using a mass spectrometer, it is never that average value that is measured. Each rainfall or snowfall has its own unique composition of deuterium and oxygen 18, which result from its history. We owe this characteristic to the fact that the physical properties of these three molecules are slightly different: the heavy molecules HDO and $H_2{}^{18}O$ have, predictably, more difficulty evaporating, but they also have a tendency to condense more easily during the formation of rain or snow.

The result of this isotopic "fractionation" is that each time little drops or snow crystals form in a cloud, there is a lessening of the amount of water vapor that created them, as well as of deuterium and oxygen 18. Gradually, as an air mass approaches polar regions and enters them, it cools, creates more precipitation, and is again weakened in terms of the heaviest isotopic molecules. So it is in the coldest region of the globe, on the Antarctic plateau, that one finds samples of snow that contain the least deuterium and oxygen 18. Thus when one goes from the coast to the center of Antarctica, the surface temperature regularly decreases. For each cooling of 1°C, the concentration of deuterium in the snow diminishes by 0.6%.[1] This characteristic is well explained by an isotopic model in which this simplified description of the life of a mass of air is translated and by isotopic models that account for the complexity of the atmosphere.[2] This is the foundation of what we call an "isotopic thermometer": the colder the temperature, the weaker the isotopic content; the same is true for the inverse (Figure 4.1).

Another such example is found in carbon 13. This time, the fractionation is linked to biological processes: the amount of carbon 13 compared to

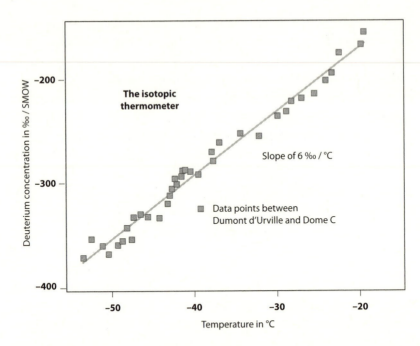

Figure 4.1. The isotopic thermometer is based on the relationship that exists, for the current period, between the average amount of deuterium and the temperature of the site of sampling. SMOW: Standard Mean Ocean Water. Source: Adapted from Claude Lorius and Liliane Merlivat, "Isotopes and Impurities in Snow and Ice," in *Proceedings of the Grenoble Symposium Aug.–Sept. 1975* (Vienna: IAHS, 1977), 125–37.

carbon 12 in the biomass is smaller than that found in the carbon dioxide of the air, CO_2, or in the ocean because during reactions of photosynthesis, plants less quickly assimilate molecules of "heavy" carbon dioxide such that carbon 12 is replaced by carbon 13. This property has many applications, including that of reconstituting ocean circulation on a global scale: as the water masses sink into the global ocean, they receive organic particles that are decomposed while forming carbon dioxide. The older a body of water, the longer it has circulated in the deep ocean, the richer it becomes in carbon dioxide and thus more depleted in carbon 13. This loss is reflected in the carbon 13 composition of the benthic foraminifers, small organisms that grow on the ocean floors using that carbon dioxide to form their skeleton.

Finally, the oxygen cycle is influenced by both biological and physical fractionations. During the process of breathing, there is preferential consumption of lighter oxygen 16; thus the oxygen remaining in the air or the water becomes enriched with oxygen 18. However, photosynthesis is not accompanied by fractionation; the isotopic composition of the oxygen produced is identical to that of oxygen present in the water used. But as we have seen, the isotopic content of that water itself depends on fractionation processes connected to physics. The signal that results, accessible from the analysis of the oxygen in the air bubbles trapped in the polar ice, is complex but very interesting from a paleoclimatic point of view.

Later we will discuss other examples of the way in which the "stable isotopes" bear witness to past climates, but these examples are the most illustrative.

Going Back in Time

As glaciologists, we pay particular attention to the glacial archives that are at the heart of this work and tell us a lot more than simply the history of the climate in polar regions. The chapters in part 2 are devoted mainly to those archives. We will first look at all the approaches that enable paleoclimatologists to describe the conditions that prevailed throughout time on the surface and in the deeper layers of the oceans, that motor of the climatic machine, and on the continents where our civilizations developed. We invite the reader to explore them by going back in time.

The Recent Period

Although some temperature records cover several centuries—the longest series is that from the center of England, which began in 1660 [3]—it was only in the second half of the nineteenth century that more systematic measuring was put into place. Gradually there were enough of these measuring networks, which were also interested in atmospheric pressure and precipitations, that mean values could be established for a given region and over the entire planet. To that end, it was important that the data be homogenized because the instruments and measuring methods were modified over time;

also many stations moved or witnessed their immediate environment change with the rhythm of urbanization. This is how Jean-Marc Moisselin[4] and his colleagues at Météo France produced a very reliable map of the evolution of the temperature of France during the twentieth century. It indicates an average warming close to one degree with slightly higher values in the southwestern quarter.

What were the conditions in the preceding centuries? Historical chronicles, popularized in France by Emmanuel Le Roy Ladurie,[5] very well documented in many European countries, as well as in China and Egypt, with series of observations that cover millennia, carry on the work. They are of a miscellaneous nature: the freezing of the canals in the Netherlands, the number of days of frost or snow, flooding and droughts, the strength of the wind and storms on the continents and oceans, the date of the migration of birds, of the blooming of trees, of grape harvests, and of the volume of harvests; these observations provide an idea of what the weather was like in those times. Even if these historical data are often of a local, largely discontinuous nature, and although they have a tendency to favor the observation of climatic extremes, they are precious sources of information. Our knowledge of the climate during the last millennia, for instance, in France, largely depends on them. The same is true for the identification of a warm period in the eleventh and twelfth centuries and of the Little Ice Age, which prevailed in Western Europe, though it was nevertheless interrupted by milder periods, from the sixteenth to the nineteenth centuries.

Data such as the dates of grape harvests, of which many chronicles give a continuous, chronological account, lend themselves to a precise estimate of summer temperatures. We then use a so-called phenomenological model that describes the evolution of the vine as a function of climatic conditions. This method has been recently applied in Burgundy, France, where dates of grape harvests have been recorded since 1370.[6] Even though years as hot as those we experienced in the 1990s had already occurred since the fourteenth century, this study confirms the exceptional nature of the summer of 2003.

We can note, however, the limits of the historical approach, even if only from a geographical point of view. Other observations—such as the spread of mountain glaciers, which is affected by temperatures and precipitation, or the temperature of the ground, which is accessible in many drilling spots and at great depth retains a certain memory of the climatic past of the region—are sources of climatic information on these timescales. Rather difficult to interpret, these data nonetheless confirm the warming of the twentieth century as compared to those that preceded it, but they do not enable us to follow its evolution in detail. Fortunately, Nature, whose development is marked by the rhythm of the seasons, offers us ancient archives, trees on the continents and coral in the oceans, which are easy to date over the years and whose growth bears witness to climatic conditions in the past.

Dendroclimatology is a discipline that examines the thickness of annual growth rings of trees as climatic indicators. It has been practiced since the beginning of the eighteenth century, with the observation, in Europe, of very small growth associated with the exceptionally cold winter of 1708–9. It is essential that the age of each ring be known precisely, and much care is given to the establishment of a firm chronology on a given site on living trees. Then, step-by-step, we identify periods showing overlaps with the help, for example, of construction wood from old buildings. Chronologies of this type enable us to go back more than a thousand years, and much longer continuous series, up to around 12,000 years, have been established from fossilized trees well preserved in riverbeds or in peat bogs. The thickness of the rings depends on the type of tree, its age, its immediate surroundings, and the availability of nutritive elements, as well as on a certain number of climatic parameters and their variation throughout the year: sun, temperature, precipitation, humidity, wind, and so forth. Climatic information can also be extracted from the density of the wood. Appropriate mathematical methods enable us to identify the climatic variables that play the most important role—for example, the temperature and precipitation during the summer—and to reconstruct climatic variations.

Dendroclimatology is nevertheless subject to calibration problems in the association of different series, and it proves above all adapted to mid and high latitudes because the trees in tropical and equatorial regions, which do not have annual growth rings, are of more marginal use.

Finally, there is an increasing number of studies on the isotopic composition of the cellulose of trees; the mechanisms of fractionation are in this case very complex, but the combined analysis of carbon 13 and oxygen 18 gives access to parameters such as temperature, relative humidity of the air, and the availability of water. Thus a study carried out on the oak trees and wooden beams in Brittany has recently revealed periods of drought over the last four hundred years.[7]

In the oceans, the coral that form many reefs between 30°N and 30°S— outside those latitudes the temperatures are too cold—also display annual growth bands. Such bands can cover several centuries—as many as eight hundred years in Bermuda. Bands result from differences in density; the densest layers are generally formed when temperatures are highest. But the composition of these corals, formed of calcium carbonate, provides additional information. Certain elements such as strontium and magnesium, which have a chemical structure similar to that of calcium, are incorporated into the calcite in the form of traces and in proportions that depend on the temperature of the ambient water. Even if other factors, such as the speed of the growth of the coral, also intervene in these concentrations, analyzing these elements is one way we can estimate the temperature at which it was formed. But the isotopic approach is probably the most fruitful. The composition of precipitate carbonate depends on the composition of oxygen 18 of the seawater and on the temperature of the water. Under certain conditions an analysis of coral can enable scientists to reach that parameter. In other cases, we attempt to reconstitute the isotopic content of the seawater, which is affected by evaporation, by strong precipitation, or even, in coastal regions, by the arrival of freshwater from rivers and streams.

The Distant Past

The advantage of dendroclimatology is that the climatic information that comes from various trees, both living and fossil, can be combined to obtain records that go back fairly far. But the approach has its limits, and it will probably be difficult to go much further than the 12,000 continuous years covered in Germany. To go further back in the past, scientists can use sedimentary layers that are deposited very slowly on the ocean floor. On the continents, one also uses sediments, deposits of loess, lake sediments, or those preserved in peat bogs, but paleoclimatology also relies on geological observations and on the analysis of matter such as stalagmites. Finally, in all cases, we must call upon other dating methods in addition to the simple counting of annual rings or bands that are so useful for trees and coral.

Paleoceanography

The oldest climatic archives are provided by the sediments that accumulate at the bottom of the oceans. In regions with little accumulation they can cover tens of millions of years, but paleoceanographers are just as interested in obtaining shorter series that, in regions where sediment accumulates quicker, enable them to describe with a great deal of detail the succession of glacial and interglacial periods that punctuated the last million years. There are many indicators that enable us to reconstruct the past functioning of oceans, their circulation, and their climatic characteristics.

Marine sediments are mainly made up of algae and the remains of small organisms, their skeletons, which are composed of either calcium (foraminifera, coccolites) or silica (radiolars, diatoms). Some foraminifera live in warm water, others prefer cold water; there are many species, and their distribution over a site provides an indication of the temperature that existed in the waters where the foraminifers were formed—on the surface for planktonic species, on the bottom for the benthics. This is true for the coccolites, the radiolars, and the diatoms; we've learned this via increasingly elaborate mathematical treatments that enable us to establish the transfer functions and to reconstitute the variations of temperature of the surface water and the deep water.

Paleoceanographers have been at the forefront of isotopic analysis. Knowing that the formation of calcite is accompanied by an isotopic fractionation that depends on the water temperature, in 1948 the Nobel Prize laureate Harold Urey proposed reconstructing this temperature from the concentration of oxygen 18 in the shells of foraminifera. Of course, this concentration depends on that of the ocean, which is variable following the period under study. Thus in a glacial period there was much more ice on the continents, 50 million additional km^3, found in large part in North America, and which contributed to the lowering of the sea level by more than 100 meters. Due to the fractionation that accompanies the formation of precipitation, these glacial ice sheets are depleted in oxygen 18 compared to the ocean. The ocean becomes more enriched with oxygen 18 when the quantity of ice on the continents is greater and when the sea level is lower. The interpretation of concentrations of oxygen 18 in foraminifera is a delicate exercise because the isotopic fractionation also depends on which species is being considered, even on its size. But by carefully choosing the sites for sampling—in places where the temperature varies little or its variation is known by another indicator—the concentrations provide information about the level of the sea in the past and thus about the rhythm of the large glaciations. These records concur very well with the variations in sea level deduced from observing the level of coral reefs that have formed "terraces," which are found in places where the coasts rise regularly due to the effects of subduction.

There is also a clear connection between the concentration of oxygen 18 of the surface ocean waters and their salinity because each of these parameters is influenced by the intensity of evaporation on the one hand and by the amount of freshwater introduced (precipitation, river water and ice sheets, melting of the sea ice) on the other. Jean-Claude Duplessy and his team at Gif-sur-Yvette developed a method from this observation that enables salinity to be measured, a parameter that proves crucial when dealing with variations in ocean circulation. We have already indicated how such variations can be documented by analyzing the carbon 13 of foraminifera; the variations in temperature can also be seen from an analysis of concentrations of certain chemical elements in the same foraminifera.

As is the case with coral, the concentration of certain chemical elements is influenced by the temperature; this is true of the proportion of magnesium to calcium, which is commonly used as an indicator of the temperature. Organic molecules are also sensitive to this parameter: some organisms respond to a change in temperature by changing the composition of their membrane at the level of molecules called alkenones, favoring nonsaturated varieties when the temperature becomes colder. The alkenones are well preserved in marine sediment, and measuring their level of saturation is another means of determining the temperature of the ocean.

Continental Archives

As we can see, the paleoceanographer is a detective on the lookout for the slightest climatic sign. The same is true of specialists of continental archives, who use almost the same range of indicators but in very different environments.

The analysis of species has an equivalent in palynology, the discipline that looks at pollens and spores produced by plants during blooming and at their dispersal in the environment. In a given site, the distribution of pollens is an indication of the ambient vegetation and thus of the climate, primarily of precipitation and temperature. These pollens are found in many sediments at the bottom of lakes, in peat bogs, and even in marine sediments of coastal regions where they have been transported by the wind. Several series cover all of the Quaternary, but most stop at the end of the last glacial period. In France, the variations in temperature and precipitation have been reconstructed from pollen series that cover the last 150,000 years in the Vosges[8] and extend over 400,000 years in the Bresse[9] and in the Velay.[10] Indicators other than pollens, such as insects and snails, can contribute to paleoclimatic reconstructions.[11]

Like marine sediments, lake sediments lend themselves to the isotopic analysis of species such as the ostracods; an analysis of their concentrations of oxygen 18 allows us to go back to that of the lake water and indirectly to that of the rains watering the catchment areas.[12] The relationship we have already mentioned between the isotopic content of the rains and the climatic conditions, more precisely the temperature in the regions under study, enables us to estimate this parameter.

Isotopic analysis is also essential to the study of limestone concretions formed in caves. The concentration of oxygen 18 in the dripping water that very slowly forms a stalagmite—here, too, through calcite precipitation—more or less reflects that of the precipitations. Thus stalagmites studied in China show the rhythm of monsoons[13] because in these regions the quantity of precipitation influences the stalagmites' isotopic content. A study done on a stalagmite from the cave of Villars in the Dordogne shows that the concentration of carbon 13 can indicate the presence or absence of vegetation above the cave, which is itself influenced by climatic conditions.[14]

Finally, climatic indications are also present in the deposits of loess, which, brought by the wind, are found in many regions. But it is more difficult to extract temperature or precipitation data from them. The same is true for many other observations: the alignment of moraines and the extent of the frozen ground in periglacial regions, retreats and advances of glaciers, the altitude at which snow remains permanently, the maximum extension of vegetation in mountain massifs, and the evolution of the lake water level in many regions of the planet.

Dating Oceanic and Continental Archives

Lake sediments can show annual layers, or varvae, which can be used for dating purposes. This can also be the case for some stalagmites, but it is exceptional in the case of marine sediments. In any event, this type of dating does not allow us to go back in time over tens or hundreds of thousands of years without interruption. Paleoclimatologists must therefore use their imagination, because what interest is there in determining the climate of a region if we don't know to what period of the past it corresponds?

Some isotopes, radioactive in this case, have proven to be of enormous help. When they disintegrate these atoms form new elements: for example, carbon 14 is transformed into nitrogen, potassium 40 into argon; this occurs at a rhythm that varies from one element to another but is well established. The initial radioactive isotope (carbon 14 or potassium 40) is characterized by the duration of its half-life, the time at the end of which half of the element will have disappeared, 5,730 years in the case of carbon 14. The concentration of carbon 14 in a piece of wood, which is measured very precisely with the help of an accelerator, will thus tell us its precise age, provided that the concentration of carbon 14 of the atmosphere at the moment when it was formed is known. Carbon 14, which is formed by the interaction between cosmic rays and nitrogen present in the upper atmosphere, has a concentration that varies little over long periods of time, but the variations linked to solar activity and to the magnetic field must be taken into consideration if we want to achieve precise dating. In practice, the inverse approach is followed: an analysis of carbon 14 in wood coming from trees, whose age we know over the last 12,000 years, enables us to calculate the initial content of carbon 14. We then can use the "calibrating" curve, thus established, to date a whole group of other materials (construction wood, organic matter, bone, shells, stalagmites, coral, and so forth), which, as we know, can be of use in archaeology and in many other domains.

Another radioactive method extends this "calibrating" curve; it is based on isotopes of uranium and thorium, the analysis of which enables us to date coral.[15] It is sufficient then to measure the carbon 14 in a well-dated coral to calculate its initial level. But beyond around 50,000 years carbon 14 dating is no longer viable because concentrations become too weak to be measured precisely. For the paleoclimatologist, the uranium-thorium method then takes over; its principal applications are the dating of ocean corals and that of stalagmites on the continental environment. This method also becomes less and less precise as one goes back further in time and is no longer useful beyond a few hundred thousand years. Methods based on the analysis of luminescence or traces of fission, both connected to the phenomenon of radioactivity, have not been found to be useful in paleoclimatology. In fact, it is astronomy that unblocked the research in this field and established a viable dating of sedimentary series covering hundreds, even millions, of years.

Beyond the fact that astronomy confirmed Milankovitch's theory, the proof of the periodicities of astronomic parameters (Figure 3.2) in marine records paved the way for a particularly efficacious dating method. Jim Hays and his colleagues obtained different records from cores drilled from the Indian Ocean. Initially they dated them while assuming a regular accumulation of sediment, which meant the oldest sample would be 450,000 years old. They then observed that the orbital periodicities characteristic of variations of eccentricity, obliquity, and precession were also present in these records and deduced from this that changes in insolation were the cause of the succession of glacial and interglacial periods in the Quaternary. This demonstration, which has been amply confirmed since then, offers, in principle, a means of absolute dating, called orbital tuning. The notion, which is applied to many types of records, consists of establishing a rough dating that is then refined by using the hypothesis that the climatic variable concerned responds with a certain delay to the variations in insolation. A study published in 1990 by Nick Shackleton, André Berger, and Dick Peltier[16] illustrates the power of this method. In a marine core, these scientists found the level corresponding to the inversion of the Earth's magnetic field, called Brunhes-Matuyama, and through orbital tuning concluded it had occurred 780,000 years ago, whereas radioactive dating based on the analysis of isotopes of argon suggested it happened 730,000 years ago. This latter estimate has since been proven wrong, and it appears that the orbital tuning estimate is correct. Furthermore, this method has been applied to all of the Quaternary and well beyond, as the variation in insolation, regardless of the period under consideration, seems to be one of the driving forces in past climatic changes, although it is not always the principal one.

A Cornucopia of Results

Ever since the first scientific observations of the climate and then the demonstration by geologists of the existence of glacial periods, there has been a considerable volume of data that have been acquired from oceanic and continental archives. This movement toward accumulating data has accelerated in the last few decades, taking advantage of increasingly sophisticated methods of analysis and dating. Our knowledge of the functioning of the climatic system has become revolutionized because of these innovations.

Even if we still have many questions regarding the implicated mechanisms, the demonstration of the link between the rhythm of the great glaciations and the position of the Earth on its orbit, which we owe to marine sediments, can be considered one of the most remarkable discoveries. An analysis of marine sediments enables us to describe not only the evolution of the sea level but also that of the circulation of currents both on the surface and deep in the oceans, providing, for example, the evidence of massive discharges of icebergs and their influence on deep water circulation. Furthermore, they offer very long-term records going back as far as 65 million years, whose analysis, as we have seen, indicates a progressive cooling of the planet, the formation of the Antarctic ice sheet, and then the more recent formation of that of Greenland.

Further back in time, most of the information is provided by clues that are currently found on the continents. Although specialists go to Namibia to look for evidence of the great glaciations from 700 million years ago, the continental archives are also very rich in information regarding those of the Quaternary and of more recent periods. The evolution of temperatures, precipitation, and drought in certain regions, of the intensity of monsoons in others, of the vegetation, of the extent of the great polar ice sheets and the glaciers, of the atmospheric circulation that was more active in glacial periods than it is today—everything, or almost everything, is accessible to whomever looks at the continental archives and knows how to decipher their complexity. These archives also hold a very important place in our knowledge, discussed in the next chapter, of the climate that has prevailed during the Holocene for approximately 10,000 years. And it is again to these archives that we turn to learn of the climate of our regions during the recent centuries and to see the imprint of human activity on it over the last several decades. For example, by looking primarily at the long series of temperatures, at the historical data, and at data obtained from tree rings, a recent study has shown that the summer of 2003 was by far the hottest in Europe in five hundred years.[17]

CHAPTER 5

Glacial Archives

In this extremely rich context, one might think that the glacial archives from those very distant polar regions are of only marginal, almost anecdotal interest. This isn't true. Although it is true that reconstructing the temperatures in Greenland or in Antarctica adds only a few more sites, the possibility of looking in detail at the evolution of the climate through the years, and over very long periods, is unequaled. Above all, glacial archives are unique because of their ability to trap the atmosphere of the past and, thanks to their extreme purity, to retain traces of the slightest effect, whether of continental, oceanic, volcanic, or extraterrestrial origin, or of that linked to human activity. Before describing the methods used by glaciologists to get a piece of ice to unveil all its secrets, we will briefly tell the long story of a snowflake from the moment it is formed to when it disappears in the ocean.

The Long Story of a Snowflake

The life expectancy of snowflakes that feed the surface of glaciers and ice sheets depends largely on climatic conditions. In cold areas without summer melting, such as the central regions of the ice sheets and the very high-altitude glaciers, they have a very long life, punctuated by several stages. Their youth varies from a few dozen to several thousand years. During this period the snow crystal rapidly loses its splendid needles, rounds out, and becomes granular, then, from the effect of more layers of snow, the grains pile up and are joined together. Nature has put into play on the surface of the white planet, well before the inventive genius of humans and on a gigantic scale, the process that powder metallurgists have perfected to create matter with satisfactory properties of cohesion and rigidity. The adolescence of a snowflake is thus the phase of transformation, called névé, which leads to its adult state:

ice, an impermeable, coherent, and more rigid material. Let's note that on the surface of temperate glaciers summer melting activates this stage of transformation and prematurely creates layers of ice as a result of the refreezing of the melting water. These layers of transparent refrozen crystals that also form in marginal and lower regions of the polar ice sheets are rather easy to identify, and their presence in greater or lesser amounts is likely to serve as a climatic indicator there.

So glaciers and ice sheets are fed through time with the snow deposited on their surfaces. The masses of ice thus formed have a tendency to grow, but their growth is regulated by different mechanisms. As we already mentioned, except at very high altitude, glaciers, as well as the more marginal and lower-altitude parts of ice sheets, are subjected to summer melting. Furthermore, ice can be deformed under the weight of the layers that accumulate over time. This plasticity of the ice is at the origin of the relatively rapid movement of glaciers and of the much slower movement of the ice of Antarctica or Greenland. The displacement of the ice is all the more pronounced when the temperature is close to the point of fusion (0°C on the surface, which lowers to −2°C under 4,000 meters of ice).

In the case of mountain glaciers, the flow follows the direction of the slope and the ablation of the ice due to summer melting increases with the loss of altitude. In some extreme conditions, the glacier can even break apart and spread into the valley, which can cause massive damage (we call this a glacier surge). The final destiny of the ice of large ice sheets is to melt either at the base under the effect of the heat emitted by the Earth at the level of the bedrock that is trapped at the interface with the ice under the effect of thermal isolation caused by the great thickness of the ice, or after floating for some time on the ocean in the form of an iceberg. The large ice sheets are calved by emissary glaciers, some of which flow straight into the ocean. In other cases, these emissary glaciers, true rivers of ice, feed the ice shelves some hundreds of meters thick and which in the ocean disintegrate, forming huge icebergs whose size, from time to time, inspires media attention. In either case, it takes a snowflake born in the center of Antarctica about a million years or more before it disappears into the ocean. In fact, the only ones that escape that ultimate fate are those that are taken by glaciologists in core samples, whose story we will now tell. But let's begin by explaining how we make this piece of ice "talk" when it arrives in our laboratories and how its age is determined.

The Ice and Its Isotopes: A Paleothermometer

The colder the temperature, the weaker the isotopic content of the snow. There is only one step between this observation, which, as we have mentioned, can indeed be verified when one travels on the surface of Antarctica and Greenland, and the interpretation of analyses of deuterium or oxygen 18 done on samples taken from an ice core at great depth. We have good reasons to think that this relationship between temperature and isotopic content is also true at a given site and that it suffices to analyze the deuterium or oxygen 18 content along an ice core to deduce the temperature at the moment when the snow was formed.

We must be cautious, however, and the best way to do this is to try to understand how water isotopes, those HDO and $H_2^{18}O$ molecules, behave between the time they evaporate on the surface of the ocean and when they arrive on the surface of polar ice sheets. To do this, we are aided by what we call isotopic models. Some are simple; they describe what occurs in a given mass of air during its travels from the ocean to polar ice caps.[1] Others are much more complex since they involve following these isotopic molecules in models of general circulation of the atmosphere, the very ones that are used to predict the temperature we will have in a few days or what the climate will be in a few decades.[2]

These models tell us that the intuition of the glaciologist, which consists of applying what he observes on the surface snow to deep ice, is justified but only under certain conditions; we will point out the two most important ones. First, it is necessary that the temperature and humidity conditions in the ocean regions from which the polar snow comes not be modified throughout time, quite simply because the conditions that rule in those "source" regions influence the amount of deuterium and oxygen 18 of the precipitation throughout the trajectory of the air masses. Second, if the proportion between the summer snow (rich in isotopes) and winter snow (poorer in isotopes) is modified in comparison to the current climate, the "isotopic thermometer" is skewed: if the proportion of summer snow has increased, it indicates a mean temperature that is warmer than it actually was.

This skewing appears to have only a marginal influence in Antarctica because the proportion between summer snows and winter snows, what we call the seasonality of precipitations, has probably not varied between the Last Glacial Maximum and the current climate, quite simply because the geography and the topography of this ice sheet have not been greatly modified between these two periods. This is not the case for Greenland, which, 20,000 years ago, was flanked by the enormous Laurentide ice sheet, whose volume was greater than that of Antarctica. The result: the seasonality of the precipitations was completely modified. It is impossible then to apply this notion of isotopic paleothermometer without complementary information; we will see that the measurement of the temperature in the bore hole and the isotopic analysis of air bubbles provide further information.

For that which concerns any distortion connected to the origin of precipitations we are also fortunate, since we know water has two isotopes with slightly different behaviors. Both can indifferently serve as paleothermometers, but in comparing them minutely we can evaluate how the "source" regions have changed throughout time and achieve a correct evaluation of the temperature at the coring site.[3] A correction, as we will see, above all necessary for Greenland.

Impurities with Multiple Sources

Apart from a few layers of ash corresponding to very intense volcanic eruptions, the observation of polar ice, to the naked eye, reveals no presence of impurities. Ice is one of the purest types of matter we know, in general on the order of a gram of impurity per ton of ice in Antarctica, more in Greenland, which is closer to continental sources. And yet the atmosphere carries dust, marine salt, and various chemical components that mix together in the snow and end up trapped in the ice. These impurities can be introduced directly into the atmosphere or be produced in it by phenomena of oxidation implied in the cycles of sulfur, nitrogen, and carbon, and those of halogen, chloride, fluoride, bromide, and iodine compounds. With their multiple sources, the impurities present in the ice, which we now know how to detect at extremely weak levels of concentration, are a unique mine of information. We will return to this, but let's first point out what we can identify in the polar ice.

Volcanic eruptions can also produce large quantities of gaseous compounds, compounds of sulfur in particular, which are transformed into sulfuric acid in the atmosphere, transported over long distances, integrated into precipitation, and, in polar regions, produce layers of more "acidic" snow, then ice. They are easily found through their electric properties, which can be systematically and continuously measured along the deep ice cores. The analysis of their chemical composition enables a description of the corresponding eruptions, some of which produce large quantities of hydrochloric or hydrofluoric acid.

In desert or semidesert regions the wind, especially when it is violent, raises dust, the finest of which can be transported as far as Antarctica or Greenland. The isotopic analysis of elements such as rubidium and neodymium makes it possible to detect their origins because the great deserts have their own unique isotopic signatures. Great forest fires also provide quantities of aerosols and chemical elements, formate, and ammonium acetate, traces of which we find in the ice of Greenland where certain pollens can also be transported, which is not the case for Antarctica. The ocean is also a great purveyor of impurities for polar ice, marine salt foremost but also sulfur compounds. These compounds are also produced by continental surfaces and by volcanoes, and here, too, isotopic analysis can be a precious help in pinpointing their origins.

These various sources—volcanic, continental, and oceanic—contribute to the complexity of the chemistry of the atmosphere, a true factory that transforms these elements but also produces them, as in the case of storms (nitrogen compounds). The chemist of ice, a *glaciochemist*, is at the end of the chain, which is an advantageous position for deciphering the atmospheric chemistry in its entirety and the way in which it has been affected by climatic changes. But he or she must first understand well the chemical processes that take place in the névé, which, in a certain number of cases—such as that in which oxygenated water is present on the surface but disappears in the depths—cause the chemical composition of the ice not to reflect exactly that of the polar atmosphere.

Human activity has added to the complexity of the chemistry of the atmosphere. We will return to this point. Traces of such activity are well noted in Greenland, which is close to anthropogenic sources of lead and other heavy metals, sulfates, nitrates, and many other substances that bear witness to human activity, and we can see in the ice that these substances have increased the oxidation capacity of the atmosphere. The Antarctic snow is not exempt from these traces of pollution: we have even found traces of radioactive elements produced during atmospheric nuclear testing in the Northern Hemisphere.

Ultimately the ice caps and ice sheets are not sheltered from what falls onto them from the sky. The interaction of cosmic rays with the upper layers of the atmosphere forms elements, which we call cosmogenic isotopes, other than carbon 14. Thus in ice we find beryllium 10, attached to aerosols, and chloride 36, in the form of hydrochloric acid; the applications based on an analysis of these elements are numerous. Antarctica and Greenland are also favored places of hunters of micrometeorites, even if they must melt enormous quantities of ice to extract usable information from it; zones with very weak accumulation are a priori the most favorable for this hunt, but nature does things well, since in certain coastal zones, where there is so-called blue ice, there is fusion then refreezing and thus a natural concentration of micrometeorites.

Air Bubbles in the Ice: A Very Beautiful Story

The initial meters of the firn are very porous, and the air circulates by convection under the effect of surface winds. Deeper down, at the bottom of the layers of firn, the air is completely isolated from exchanges with the exterior and is then imprisoned in the ice in the form of bubbles (Figure 5.1). This process takes a few dozen years for relatively warm sites with high accumulation and as many as several thousand years for those in the central regions of Antarctica. It is within this range of time that the air bubbles are younger than the ice that encloses them.

After the bubbles are enclosed they become smaller and the air pressure increases with depth. As we go deeper, the bubbles gradually disappear, to the point of no longer being observable under several hundred meters of ice: air molecules are then imprisoned inside the ice matrix in the form of air hydrates (also called clathrates). The ice, without its bubbles, becomes perfectly translucent, even though the volume and the composition of the air it encloses are unchanged in relation to the initial bubbles.

It is an old idea, perhaps first expressed by Per Scholander in the 1950s, to try to extract the air from the past from these little bubbles that escape by the thousands when one plunges a "polar" ice cube into one's whiskey (an obser-

Figure 5.1. Air bubbles trapped in the ice. © CNRS Photothèque/Volodia Lipenkov.

vation that was at the origin of our quest for fossil air that began in the 1960s!). In the 1960s and 1970s two teams, one from France, directed by Claude Lorius, and the other from Switzerland, directed at the time by Hans Oeschger, developed analytical methods to reconstitute the past variations in the concentration of carbon dioxide in the atmosphere from air bubbles trapped in polar ice. The teams' objectives were to extend back in time the carbon dioxide record initiated by Charles Keeling a few years earlier from atmospheric measurements and to explore the hypothesis proposed seventy years earlier by the Swedish scientist Svante Arrhenius regarding the origin of glacial periods. This story deserves to be told.

In April 1896 Arrhenius, a physics professor on the faculty of Stockholm University, a private scientific institution, published an article titled "On the Influence of Carbonic Acid in the Air upon the Temperature of the Ground" in the *London, Edinburgh, and Dublin Philosophical Magazine and Journal of Science*. This work was truly important, but it was overlooked for a long time in the study of what today is known as the "greenhouse effect." It was the Frenchman Joseph Fourier who, in his "Remarques générales sur les températures du globe terrestre et des espaces planétaires," presented in 1824 in the annals of the Académie royale des sciences of the Institut de France, was the pioneer of this concept, which constitutes a major challenge for the future of the environment of our planet. He spoke at that time of the luminous heat received from the Sun by the surface of the Earth and of the obscure heat reflected by the Earth, which, in going back through the atmosphere with more difficulty, contributes to global warming. But it was Arrhenius who was the first to evaluate the influence of a change of atmospheric content in carbon dioxide on the surface temperature of the Earth. How can we not admire him when we know that the calculations carried out a bit more than a century ago by that great scientist—he received the Nobel Prize in 1903, but for his work in chemistry—led to an estimate of the warming of the Earth's surface similar to what the community of researchers has provided at the beginning of the twenty-first century with the help of computers, which are among the most powerful available! Luck, genius . . . . how can we explain this coincidence?

Arrhenius had a career as a physicist-chemist behind him when, at the beginning of the 1890s, he founded the Stockholm Physics Society, which quickly became a mecca for interdisciplinary exchanges. It was there that in 1893 he attended a lecture by Nils Elkohlm from the Swedish Meteorology Office on the origin of glaciations. A discussion took place during which the participants granted only little credit to the influence of the variations of the Earth's orbit or to that of the displacement of the poles on the surface of the planet. Then, the following year, he followed attended a lecture given by one of his colleagues, Arvid Högbom, on the reasons why the content of atmospheric carbon dioxide could vary throughout geological periods. It was probably through the effect of these two debates that Arrhenius undertook what he called "tedious calculations" regarding the causes of the existence of the ice ages, with the idea of linking climatic changes to those of CO_2 over the long term. In one year—a surprisingly short amount of time—he conceptually developed the first approach based on empirical data to what today is called the modeling of the evolution of the climate in response to a change in atmospheric CO_2. He thus calculated what must have been the variation in average temperature depending on the season and the latitude for various contents of atmospheric CO_2.

The results of these "tedious calculations" were published in April 1896. Arrhenius's primary motivation was to demonstrate that the cooling that corresponded to the beginning of ice ages could be explained by a decrease on the order of 40% of the ability of CO_2 to absorb the "obscure heat" of Joseph Fourier. And thus to refute Croll's astronomical theory, which proposed a different angle.

We can understand glaciologists' motivation to attempt to determine if such a decrease in atmospheric CO_2 had indeed taken place during the glacial periods. It was nevertheless necessary to wait until the beginning of the 1980s for our quite natural curiosity to be satisfied. Since then specialists of air bubbles have worked extremely hard on the CO_2 record during the past and on other greenhouse effect gases—methane, CH_4, and nitrogen protoxide, N_2O—which are also influenced by human activity. The analysis of carbon 13 in carbon dioxide, but also of deuterium in methane, later proved capable of helping us understand the origin of the variations of the compounds and the way in which their natural cycles function or are disturbed by human activity. These scientists didn't stop there; they began to explore the isotopic compositions of other molecules present in the air, in particular those of its primary components, nitrogen, oxygen, and argon.

Let's take the case of nitrogen and argon, whose isotopic compositions we know have remained constant in the atmosphere over millions of years. We note, nevertheless, that they vary in air bubbles. Jeff Severinghaus, an American researcher, has even shown that these changes were abrupt at the time when the temperature of sites with high snow accumulation rates altered rapidly.[4] These variations are explained by the combination of isotopic fractionations, one linked to gravity—there are slightly larger quantities of heavy molecules at the bottom of the névé—and the second to thermal origin, which results from the difference in temperature between the surface of the ice sheet and the base of the firn. Consequently, the analysis of nitrogen and argon isotopes offers a method, rather surprising but precise and thus widely used, to estimate changes in temperature when they were relatively rapid due to the effect of thermal fractionation. Temperature measurements carried out in the bore holes of Greenland had shown that the estimates deduced from the composition in oxygen 18 of the ice underestimated by half the cooling of the Last Glacial Maximum there. The analysis of air bubbles has since shown that such an underestimation there, linked to the change in seasonality of the snowfalls, was systematic throughout the last glacial period.

The oxygen 18 in the air trapped in the ice is modified by these fractionations, but the greatest variations observed have a completely different origin.[5] They reflect both the changes in isotopic composition of the seawater, which essentially depends on the quantity of ice amassed on the continents, and the modifications of the continental biosphere and the hydrological cycle in the low latitudes strongly affected by variations in insolation. One can be free of the influence of the latter and thus have access to past variations in sea level.

Still looking at the air but now just the amount of it and not its composition, the idea is simple: the air pressure at the moment the bubbles were enclosed, and thus the quantity trapped, decreases with altitude; in principle it would suffice to measure the volume of air trapped to know how the altitude of glacial ice caps has varied over time. The use of this paleoaltimeter over several coring sites, coupled with the modeling of the flow of ice, has revealed that during glacial periods the ice on high plateaus of the ice sheets was thinner due to lesser snowfall, whereas the regions more on the periphery had thicker ice and a lower sea level, which enabled the ice cap to extend farther. Putting this simple idea into play, however, became unexpectedly difficult. Surprisingly, the volume of air also appears to depend on insolation, whose variation may influence the metamorphism of the snow during the formation of the firn and thus the mechanism of trapping, processes that also seem to affect the proportion of oxygen to nitrogen. This is very interesting from the point of view of dating.[6]

The Headaches of Dating

Compared to the dating of continental and oceanic archives, ice dating has its own specific issues. The first comes from the fact that some of the information we are interested in is inscribed in the ice, some in the air bubbles. Because the air is younger than the ice—since air trapping occurs gradually, these ages are averages—we need two distinct chronologies. Further, ice cores hardly lend themselves to the use of radioactive methods. Carbon 14 dating is generally less precise than other methods for dating ice cores. It can be used in exceptional cases, for example, if there is vegetal debris present—

this is the case in a core from the Andes. In addition, carbon 14 dating is inapplicable beyond a few dozen thousand years.

What methods, then, do glaciologists use to establish the chronology of ice core samples, the oldest of which, on the Antarctic site of Dôme C, spans 800,000 years? Here, too, the simplest solution involves counting annual layers, because many of the properties of the snow—concentrations of deuterium and oxygen 18, electrical conductivity, which depends on the content in impurities, composition of many chemical elements—differs depending on whether they accumulated in the summer or the winter. Furthermore, in the summer dust particles are more abundant in the snow because the winds are more favorable at that time to their being carried to the poles: the corresponding layers diffuse more light. The visual observation of the successive layers of snow, on the walls of pits several meters deep dug for the purpose of counting and then dating the recent layers of snow, has been used a great deal, and the stratigraphic variations sometimes remain visible at a great depth. None of the indicators is perfect, but taken together they enable an annual dating as long as the thickness of the layers, which become thinner as we dig into the ice sheet, remains sufficient. Annual dating has been possible in the coastal regions of Antarctica, and more especially in Greenland where these indicators have been applied to date ice as old as tens of thousands of years.

Some events can, if they are dated in other ways, help establish the chronology of ice core samples and validate a dating obtained by counting layers. This is true of the generally unambiguous volcanic eruptions that we can detect by analyzing physical properties, such as the electric or chemical conductivity (marking the presence of sulfates) of the ice. This also true of certain deposits characteristic of ash layers.

Recent large, well-documented eruptions, combined with spikes in radioactive fallout elements produced during nuclear arms testing in the 1950s and 1960s, are useful landmarks. Further back in time, the process is reversed: thanks to precisely dated coring in Greenland, the chronology of volcanic eruptions during the Holocene has been established. Even further back, major eruptions whose dates we know within a few thousand years using radiometric methods, such as that of Toba, which occurred around 75,000 years ago, are used to verify the chronology of deep ice cores. Similarly, we can use a spike in the cosmogenic isotopes fallout, beryllium 10[7] and chloride 36, from around 40,000 years ago. According to the most wide-spread hypothesis, this spike would have a connection with the Laschamps event, which corresponds to an abrupt decrease in the geomagnetic field.

The analysis of beryllium 10 offers other possibilities. The production of this isotope depends on the intensity of cosmic rays and on that of the geo-magnetic field, but it also varies with solar activity, a factor that seems domi-nant for the Holocene. These mechanisms are identical to those that rule the production of carbon 14. Even if the processes of deposit differ, there are re-markable similarities between the variations of carbon 14, deduced, for ex-ample, from the analysis of tree rings, and those of beryllium 10 in Antarctic ice. Grant Raisbeck and Françoise Yiou, researchers at CNRS in Orsay, have thus established an absolute chronology of Antarctic ice over the last 7,000 years.[8]

Going back further, glaciologists have developed models that enable them to calculate the thickness of successive annual layers. These layers become gradually thinner as one goes deeper because of the effect of the flow of the ice, and their thickness is equal to the initial value multiplied by the thin-ning. The establishment of glaciological dating associates a flow model that provides thinning and a model of the history of surface accumulation that varies depending on the climate. It is weaker in glacial periods, as masses of colder air that bring precipitation over the ice caps then contain less water vapor. The age is calculated by counting the number of years that have elapsed between two successive depth levels, then by adding them starting from the surface. This method has been applied to all deep ice cores in Antarctica. It is also useful to estimate the difference in age between the ice and the trapped air. The depth at which the pores close up and trap the atmospheric air is all the greater when it is cold and the accumulation is great. The model devel-oped by Jean-Marc Barnola in Grenoble and his team calculates this depth and the associated difference in age; in a glacial period it can reach 8,000 years.

These glaciological models have the disadvantage of being less precise as the age increases, and it is then indispensable to have points of reference (often called pinpoints) deduced from comparison with ocean records, themselves dated by orbital adjustment. It might appear more reasonable to achieve this orbital adjustment over variables recorded in the ice cores them-selves. Periods linked to the Earth's orbit are visible, at least in the sampling of

Antarctica, when one analyzes the concentration of deuterium in the ice, those of oxygen 18 and methane in the air bubbles, and, as we've mentioned, the volume of air trapped and the relationship of oxygen to nitrogen in the air bubbles. Thus, the oxygen 18 in the air, studied by Michael Bender's team from Princeton University, varies with a periodicity of around 20,000 years, which is directly connected to precession.[9] However, these orbital chronologies generally have the disadvantage of being based on the hypothesis of a constant interval—we speak of constant phase relationship—between the considered property and the variations in insolation, a hypothesis whose validity has not been established. Conversely, the mechanisms that we qualify as nonlinear are certainly in play when the climatic system is complex, mechanisms that are translated by the variable phase relationship between the insolation and one of those parameters, which these orbital chronologies do not take into account.

This type of problem is raised in many domains, notably for geophysicists who have developed "inverse" methods that prove well adapted for dating ice core samples. In that case, it is a matter of associating the information provided by several dating methods by taking into account the uncertainty surrounding each of them. This approach, taken by Frédéric Parrenin, consists of adjusting, with the help of numerical processes, the values of the parameters of the glaciological model in such a way that once it is fixed, the dating obtained agrees as much as possible with all the available chronological information.[10] This inverse method, which has the advantage of not imposing a constant phase relationship, is no doubt the best adapted and the most flexible in the absence of observed seasonal variations that alone enable a year-to-year dating. Nevertheless, the ages it provides are tainted by a relatively large uncertainty—a few thousand years—as soon as one moves away from the period for which well-dated layers exist. This inverse method has been adopted for large-scale deep ice core records in Antarctica where the accumulation is very small, whereas the counting of layers remains the favored approach in Greenland.

New approaches are emerging that will allow us to achieve an absolute dating of long records of glacial-interglacial cycles: the volume of air contained in the bubbles, a parameter that we have already mentioned, and the oxygen-nitrogen ratio, whose very weak variations appear to have a connection with the variations in summer insolation at the drilling site.[11] Although the physical processes involved need to be confirmed, these two parameters are promising markers that record summer variations at the site where ice was formed, variations whose age we can know with great precision using astronomical calculations.

We will now look at the detailed history of these deep ice cores.

CHAPTER 6

The Campaigns

Copenhagen, July 22, 1952. Willi Dansgaard carefully collects the rain produced by a storm occurring over the capital of Denmark. He is interested in the oxygen 18 composition of the successive samples. If it is true, as we learned in school, that the formula for water is written H_2O, that which we drink, as we have mentioned, contains different isotopic molecules, $H_2^{16}O$, $H_2^{18}O$, HDO, and so forth. The different isotopes of water were identified in the period between the two world wars, and the variations in their concentration in a natural environment were quickly proven. But Dansgaard's study opened the path to a systematic exploration of the distribution of water isotopes in precipitations. The article he published in 1964 on his findings,[1] and which is the study's high point, is still often cited. One of the remarkable aspects it reveals is the relationship between the temperature at the site and the isotopic composition of precipitations. The colder it is, the weaker the number of $H_2^{18}O$ molecules as compared to those of $H_2^{16}O$, and inversely. The variations are minor but largely sufficient to be detected thanks to the development of mass spectrometry, which highlights the fact that the $H_2^{18}O$ molecules (mass 20) are heavier than those of $H_2^{16}O$ (mass 18). With the help of a simple model Dansgaard explains this observation, but more important, he was one of the first to measure its potential applications, which logically must give access to temperatures that existed in the past, as long as it is possible to have access to past precipitations.

Camps Century and Byrd: The First Deep Ice Core Drillings

The idea was simple, and it was natural for Dansgaard to propose that it should be applied to Greenland, a Danish land by tradition, but today an autonomous territory. That ice sheet, more than three kilometers thick in the central regions, contains ice that is increasingly old as one digs into it. But digging into the ice—obtaining samples—is no small feat. Fortunately the icebergs that calve around Greenland offer an alternative solution because they are formed by more or less ancient ice depending on whether they come off the center or the edges of the ice cap. A campaign led in 1958 by Willi Dansgaard enabled the sampling of an entire series of icebergs and the dating of them, thanks to the content of carbon 14 in the air that was extracted from them. The oldest sample, approximately 3,000 years old, was formed in the central regions, and its very weak content of oxygen 18 indeed reflects the very cold temperatures that existed there related to high altitude. A French team, led by Étienne Roth at the CEA Saclay, was also very interested in those icebergs. This team was in charge of checking the concentration in deuterium during the enrichment processes leading to the production of heavy water, D_2O, and thus developed methods to analyze the content of deuterium, which it proposed to apply to the Greenland icebergs.[2] The result they obtained wasn't surprising. As anticipated from the fact that fractionation processes are similar for HDO and $H_2^{18}O$, the content of deuterium and oxygen 18 varied in parallel, but thanks to this study the French geochemists were henceforth involved in the polar adventure.

There was not a lot of hope, however, of going back to the distant past using icebergs. The only means to do so was to dig into the ice sheet, and this objective was made possible thanks to support from across the Atlantic. In Hanover, New Hampshire, a laboratory of the U.S. Army, the Cold Regions Research and Engineering Laboratory (CRREL), was dedicated to the study of cold regions. Polar ice caps, snow-covered surfaces, permafrost, and sea ice were the main areas of interest of this remarkable lab. One of its teams was involved in the development of core drills, instruments that enabled glaciologists to extract cylinders of ice a dozen centimeters in diameter and one to three meters long and to go back through successive "stages" to increasingly deep ice. Further, the CRREL had access to large military transport planes, C130s, which were able to transport from the U.S. East Coast large quantities of materials to Greenland. There, about a hundred miles from Thule, the military base Camp Century (Figure 1.1) was established by the U.S. Army Corps of Engineers in 1959, a true city under the ice with roads, houses, a movie theater, a church, and a network of trenches and tunnels, the longest of which, Main Street, stretches along 350 meters. Everything was powered by nuclear energy.

For obvious logistical reasons, this was the site chosen for the first deep ice core drilling in a polar region. The operation in which the Danish team and that of the CRREL, led by Chet Langway, participated was a success. The rock base was reached at a depth of 1,390 meters in July 1966—not without difficulty, however, as it took six successive seasons to achieve that depth. New methods for analyzing the physical characteristics of the ice, its electrical conductivity, and the impurities that it contained were put into place. But it was the analysis of the oxygen 18 of this ice, undertaken in Copenhagen, that delivered the clearest message. The bottom of the drilling around 1,300 meters had less oxygen 18 than the first 1,100 meters, and these results were beyond doubt. This was ice from the last glacial period, whereas the upper part was formed during the Holocene, in the last 10,000 years or so. A glaciological model enables scientists to estimate the age of the ice accounting for the gradual thinning of the layers as they become deeper. Whereas the first 1,000 meters represent around 6,000 years, the last 200 cover around 100,000. All of the last glacial period is indeed there but increasingly packed down the closer one gets to the rock base. The interpretation of the results was not simple, but that year, 1966, was momentous. With Camp Century, glaciologists had the first deep core drilling to use to reconstitute our climate's past.[3] It was just in time because the following year Camp Century had to be closed due to relatively rapid movement of the ice, which made it impossible to maintain the camp. Fortunately the nuclear plant was moved in time.

Confident from this success, the drillers of the CRREL went on to Antarctica. There, too, the drilling site, at Byrd Station (Figure 1.2) on the ice cap of West Antarctica, was chosen because of its accessibility from the American McMurdo Base. Nearly 2,000 people live in this true town during the summer and a few hundred the rest of the year. It is located at the base of the Ross Sea a few hundred kilometers from the South Pole, where the United States has constructed a permanent base. It was near McMurdo that, at the beginning of last century, various expeditions started off in the challenging conquest of the South Pole, this somewhat mythical place. Visiting the huts of Scott and Shackleton, which have remained as they were originally, is a very moving moment for anyone who has the opportunity to stay a few days at McMurdo. Byrd Station is located a few hundred kilometers away in a place where the ice cap is a bit more than two kilometers thick. There, too, the rock base was reached in 1968 after only two seasons. This was a great success, even if the age of the ice was discovered to be no more than 80,000 years old.[4] Unfortunately the drill remained at the bottom of the hole and we had to wait until 1993 for American drillers to again celebrate the success of a deep drilling.

Another team then joined the Danish and American pioneers, that of the physicist Hans Oeschger from the University of Bern, to whom we owe the perfection of counters adapted to precisely measure very low-level radioactivity of elements such as tritium and carbon 14. The analysis of tritium was interesting for following the fallout due to nuclear explosions on the ice sheets and for dating the recent layers of snow,[5] whereas that of carbon 14 had to be adapted to the dating of older layers. Oeschger and his laboratory were primarily interested in Greenland, where they participated in various campaigns aimed at obtaining recent samples and then brought their contribution to the Byrd program. Enthralled by the complexity of how the climatic system functions and by environmental issues in general, Oeschger made his mark on the relatively recent development of those disciplines in which the Bern lab, now directed by Thomas Stocker, is still very active.

The research carried out by glaciologists in polar regions in the 1950s and 1960s was not limited to these Danish, American, and Swiss contributions. Other countries, Scandinavia in particular, have been active in Greenland. For Antarctica, the 1957–1958 International Geophysical Year was the point of departure for research activities in which a dozen countries were involved. In 1959, in a spirit of collaboration, they signed the Antarctic Treaty, which devoted the continent to peace and research.

The properties and the composition of the snow taken from the surface and from shallow drillings were studied. That was the opportunity for an American team to determine the isotopic content of the snow at the South Pole.[6] There, too, a relationship with the temperature is very marked. As we might expect, the snow has fewer heavy isotopes (oxygen 18 and deuterium) during the winter than it has during the summer. In the next section we'll discover what the French teams did in Antarctica and Greenland in the 1950s and 1960s.

Fifty Years Ago: The French on the Polar Ice

After the pioneering work of Jules Dumont d'Urville in Antarctica and Jean-Baptiste Charcot in Antarctica and Greenland, French interest in polar regions was renewed through Paul-Émile Victor and his expeditions to Greenland. After his winter stay devoted to the study of the Inuits, he founded the Expéditions Polaires Françaises (EPF) in 1947, which, until 1992, supported scientific missions to the two poles. From 1957 to 1960 he directed the International Glaciological Expedition, which was responsible for studying the Greenland ice sheet following a west-east axis and for installing a wintering station in its center. French researchers participated in this work, but at that time the national research effort was aimed primarily at Antarctica, with the beginning of the International Geophysical Year (IGY) in 1957, which saw twelve countries invest in the study of that vast continent, which was still practically unknown. That IGY produced a wealth of scientific results. Under the aegis of the French Academy of Sciences, research undertaken in Adélie Land benefited from the establishment of the coastal Dumont d'Urville Base and truly took off thanks to three expeditions that occupied this base and that of Charcot (see the drawing on page 95), installed on the glacial ice sheet 320 kilometers to the south, at 2,400 meters in altitude. The climate was only one of the areas studied, and the idea that the polar ice sheets contained unique archives had not gained much support at that time. French glaciologists measured the thickness of the ice with the help of seismic probes, as well as the accumulation of snow and its properties on the surface from shallow drilling, which also enabled them to determine the mean temperatures by placing a thermometer at the bottom of the holes they drilled.

In *Les glaces de l'Antarctique*,[7] Claude Lorius passionately recounts the adventure of the year he lived in Antarctica in the company of Roland Schlich and Jacques Dubois. From an explorer at the beginning in the 1960s he was transformed into a geochemist. Convinced of the power of the isotopes of water as a tool for reconstituting the climates of polar regions, he devoted his doctoral thesis to this; his primary objective was documenting their distribution in recent snow. He participated in fieldwork that went back and forth over Victoria Land and Adélie Land and took many samples from both areas. At the end of the 1960s, two drillings down to around one hundred meters were undertaken in the coastal regions by a French team.

But the scientific tools of the EPF were relatively rudimentary. A barge docked on the quays of the Seine served as a laboratory, and it did not have such a sophisticated apparatus as a mass spectrometer. Fortunately the research center of the CEA in Saclay was only about twenty kilometers away. There then began a successful collaboration between isotopists from the CEA and glaciologists, which has produced a wealth of results during almost fifty years of work together. In the CEA, the Laboratoire de Géochimie Isotopique (LGI) directed by Liliane Merlivat within the department of Étienne Roth has been the backbone of the operation. Deuterium mass spectrometers with a unique system of automatic injection were developed at the CEA. The content of that isotope was thus analyzed there in preference over that of oxygen 18. A very strong correlation between deuterium and the mean average temperature of the site was established over the entire zone explored by Claude Lorius. Isotope and temperature varied hand in hand: a cooling of 1°C was accompanied by a decrease in the amount of deuterium by 0.6%.[8] This ratio, established in the 1970s, is still the one used to interpret deep drilling carried out in that sector of East Antarctica.

Dominique Raynaud was the first Ph.D. student of Claude Lorius, who encouraged him to work on a new and promising subject: the composition of the air trapped in polar ice. He started in the 1960s working at the CEA and looking at the CO_2 content of the ice. It was a pioneering work, but at that time the methods used for extracting the air from the ice provided inconclusive CO_2 results. Finally Raynaud formulated a thesis around a simple idea. Since air pressure decreases the higher one goes in the atmosphere, the amount trapped in the ice must decrease as the altitude of the ice sheet rises. Even if the interpretation of the results is more complicated than it first appears, Raynaud demonstrated that 20,000 years ago, in the Last Glacial Maximum, the glacial ice sheet was thicker than it is now, probably on the order of one kilometer at Camp Century.[9]

In 1968 Jean Jouzel was just beginning his doctoral thesis at the LGI. A research topic off the beaten path was proposed to him by Étienne Roth: What can we learn about the formation mechanisms of large hail from their isotopic analysis?[10] This was at a time when much hope was placed in the possibility of preventing devastating hail by injecting clouds with silver iodide. Basic research on hail went along with these more applied programs. They were, however, almost abandoned a few years later when rigorous tests of prevention methods proved to be ineffective. But it wasn't far from the ice of hail to that of glaciers, and Jouzel enthusiastically made the leap.

The barge at Saint-Cloud was not big enough for the research team that Claude Lorius hoped to create around his research in polar regions. With the speedboats and other barges traveling on the Seine River, the barge was also not stable enough to establish an analytical lab there! Jacques Labeyrie invited the team to the Centre des faibles radioactivités in Gif-sur-Yvette. The offer was interesting, but Lorius was attracted by the Laboratoire de Glaciologie in Grenoble whose director, Louis Lliboutry, was a brilliant glaciologist. Excited by the modeling of Alpine glaciers, he was less interested in the study of polar ice caps as climatic archives. But he was ready for his laboratory to be involved in it. The goal was quickly established: Lliboutry agreed that Lorius should take the lead on a project to carry out deep drilling at the site of Dôme C, in the heart of the Antarctic continent, at 1,100 kilometers from the Dumont d'Urville Station (Figure 1.2). Unlike Camp Century and Byrd Station, the site was deliberately chosen. The snow that falls on a dome comes down vertically, and, providing that the dome did not move significantly over time, the ice taken from along the length of a core sample is, regardless of the depth, formed from snow that, to a few kilometers, has fallen on the site itself. Thus the interpretation of the results was to be largely facilitated with respect to a site where ice at depth is originated from upflow.

The First Drilling at Dôme C: Success of the French Team

The project began with the development of a core drill. There are essentially two systems involved in deep core drilling: a thermal core drill and an electromagnetical core drill. The thermal core drill, equipped on its end with a heating mechanism, penetrates through fusion a thin circle of ice while that circle is cut from the rotation of knives in the electromechanical core drill. In both cases, the ice cores are extracted in a cylindrical core tube and brought up to the surface at the end of each drilling. This tube is equipped with a reservoir that allows debris to be brought up, as well as liquid water and ice chips. A motor, pumps, and associated electronics are all connected to the core drill, which is suspended by a cable that provides energy and signals from the surface where the team of drillers is installed next to a winch, a drilling tower, and a command console. Many tests are necessary before everything is operational, and there is always the risk that either core drill will become blocked and not be able to be brought back up. At Camp Century, the CRREL team had two failures with a thermal core drill. On the third try they used that system at the beginning of the drilling, but at around 600 meters went to an electromechanical drill that allowed them to reach the bedrock. This second system was also used at Byrd, but the decision to drill very long six-meter cores called for an extremely cumbersome system. With a 26-meter-long tube and a winch of close to 20 tons, the apparatus weighed nearly 70 tons.

In Grenoble, the team of François Gillet opted for a thermal core drill. The idea was to achieve a dry drilling without any filling fluid, which would enable them to avoid the closing of the hole. This reduced the amount of material that needed to be transported to the site but forced them to carry out the drilling in a single season. There was little hope, in the end, of going beyond 1,000 meters because in the absence of fluid the drilling hole could quickly become deformed. In the spirit of cooperation that ruled in Antarctica, McMurdo provided logistical support in the form of C130s that the U.S. Army put at the disposal of the National Science Foundation (NSF). Two planes crashed in 1975 during the first season of reconnaissance, both of which were near tragedies. Despite these setbacks, the NSF maintained its support of the French drilling operation at Dôme C, and three years later there was success; in less than two months the drilling team, under the direction of Daniel Donnou, reached a depth slightly greater than 900 meters. Seven tons of ice were brought back to our labs. We estimated that the deepest ice was about 30,000 years old[11] but later revised that dating and now more than 40,000 years of archives are available from the first drilling at Dôme C.

This was a true godsend for the French teams from Grenoble, Saclay, and Orsay. After several years of research Dominique Raynaud and Robert Delmas, a transfer from Saclay to Grenoble, perfected a method for extracting air bubbles trapped in the ice, which enabled analysis of their concentrations of carbon dioxide. This was also the main objective of Hans Oeschger and the glaciological team of Bernhard Stauffer in Bern. The Swiss team at Byrd and the French at Dôme C had access to ice from the Last Glacial Maximum, and both teams contributed to a major discovery and confirmed the prediction made by S. Arrhenius at the end of the nineteenth century: the concentration of carbon dioxide was indeed at that time about 30% less than that of the preindustrial period before human activity began to change it.[12]

The content of dust,[13] the size of crystals,[14] and the chemistry of the ice[15] were realms in which the scientists from Grenoble, under the leadership of Martine de Angelis, Robert Delmas, Paul Duval, Michel Legrand, and Jean-Robert Petit, were to distinguish themselves, thanks to the ice from Dôme C. At Saclay the emphasis was placed on the conjoined analysis of the two isotopes, deuterium and oxygen 18. One or the other can provide access to the temperature, but, on largely theoretical bases, one could expect that a detailed comparison of these two isotopes would provide supplementary information. The wager was won. The comparison gave access to the conditions that prevailed in the ocean regions from which the air masses that generated precipitations came.[16] Consequently, the conjoined analysis of deuterium and oxygen 18 has been henceforth systematically carried out on the ice of Antarctica and Greenland. At Orsay, Françoise Yiou and Grant Raisbeck demonstrated that it is possible to measure the concentration of beryllium 10 in polar ice, and analyses done on the ice of Dôme C suggest that there is a way to reconstruct the past variations in the accumulation of snow.[17] With this wealth of results, the French entered the very tight circle of nations that had succeeded in deep core drilling in polar regions.

We are proud of these initial successes. But in this realm of research dedicated to the reconstitution and understanding of past variations in the climate, the high ground was indisputably held by paleoceanographers, whose strength lies in their access to data on variations in sea level provided by the analysis of oxygen 18 in foraminifera. The distribution of the species that find their optimal growth temperature in more or less warm waters further provides access to variations in temperature. And above all, marine sediments enable us to go far back in time; many marine cores cover several climatic cycles of the Quaternary and even beyond it. At the time of this first Dôme C record, we talked in our scientific community about the works already mentioned of Jim Hays, John Imbrie, and Nick Shackleton, which had just established the validity of Milankovitch's theory. Much remained to be clarified regarding the mechanisms put into play, but to make a contribution glaciologists should necessarily go further back in time.

Rapid Climate Variations: Initial Inklings

The Danish and American teams were also attempting to go back in time. With the loss of the core drill at Byrd, the drilling team from CRREL was unable to continue their work. The Copenhagen team took over. Engineers, technicians, and scientists all worked under the direction of Niels Gunderstrup and Sigfus Johnsen. The electromechanical drill Istuk was built and tested during the 1970s. This new drill had a motor driven by rechargeable batteries installed in the drilling tube. Built to drill cores two meters long, it was much less cumbersome and lighter than the one used at Byrd; it was 11.5 meters long and weighed only 180 kilograms. The winch and the drilling tower were scarcely more than a ton because the cable was not very thick.

Despite these efforts to construct a relatively small, lightweight drilling system, there were logistical constraints that dictated the choice of site for this new drilling project, GISP (Greenland Ice Sheet Project), which the Swiss from Bern joined, along with the Danes and the Americans. The scientists argued in favor of drilling at the center of Greenland at its highest point, convinced that they would reach the oldest ice there. But that region was far from the American bases, Thule to the northwest and Sonderstrom Fjord to the southwest. A site close to the latter, Dye3 in southern Greenland was ultimately chosen, even though it was far from ideal. The origin of the ice formed upstream from the site was difficult to determine because this coastal region is very hilly. It is also a relatively warm site where summer fusion prevents the reconstruction of variations in the composition of air bubbles, which are very sensitive to fusion. After three seasons (1979–81), the bedrock was reached at a depth of 2,038 meters. The scientists' fears proved grounded, as the samples did not go beyond the last glacial period, and their analysis was complicated. But this drilling offered an extremely surprising discovery: that glacial period and the deglaciation that followed it were marked by great and very rapid warming that occurred in the span of a few decades, or perhaps even a shorter period of time. These rapid variations were also seen at Camp Century, but these results were not put forward because the scientists feared that the variations were due to the movements of the ice near the base. Documenting them at a second site eliminated any doubt.[18]

These rapid variations, henceforth known as Dansgaard-Oeschger events, were indeed of climatic origin.

Vostok: A Collaboration between French and Soviet Teams

To go back in time is, a priori, more promising in Antarctica than in Greenland. The accumulation of snow there is weak, less than 10 centimeters per year, or around 3.5 centimeters of water, over a large part of the Antarctic plateau, and the ice is thick, often more than three kilometers. In addition, the movement of the ice is very slow throughout the central region. The flipside: accessing it is not easy and the temperature is the coldest on our planet, making fieldwork much more difficult. In spite of everything, the Soviets established a permanent station in this central region at the Vostok site (see Figure 1.2), where the record for the lowest winter temperature ever was recorded (−89.2°C). At the beginning of the 1970s a long-term operation was begun at Vostok—a true epic adventure under the direction of the Leningrad Mining Institute, whose drillers were accustomed to extreme conditions—which, after thirty years, has just been completed.

Les glaces de l'Antarctique tells the first part of the story of Vostok. It had nothing in common with Camp Century, Byrd, Dôme C, or Dye3 where the drillings were carried out during the summer months. At Vostok, drilling was done throughout the year, and many ice cores were extracted in temperatures lower than −70°C. Living conditions were precarious, and supplies came only in the summer overland from Mirny Station. The Russian drillers alternated the use of thermal and electromechanical drills, and alcohol was commonly used as an additive to give the kerosene used as a drilling liquid an appropriate density. The Russians are uncontested specialists in deviation: when a core drill is blocked, they have to bring up the cable by some fifty meters, block the drill hole, then dig some more by deviating the drill, which enables them to begin another hole, which is then carried out a few meters from the first without starting from the surface again (Figure 6.1). Nevertheless, they didn't escape the harsh reality of glacial drilling: stuck drills, lost cable, and so forth. The only solution in cases such as those was to begin again and to start at zero from the surface. At Vostok, the first drilling reached 500 meters in 1970. The second was completed at 950 meters in 1972. The third seemed to be the best, 1,400 meters in 1980, 2,083 meters on April 11,

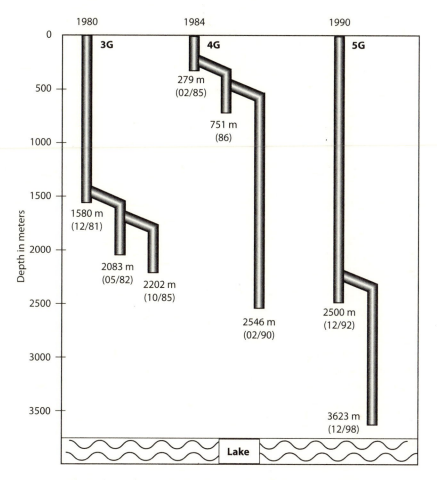

Figure 6.1. Chronology of ice core drilling undertaken at Vostok since 1980.

1982 (Figure 6.1). The next day, at 4:00 A.M., the station's electric generator caught fire. Operations were suspended because the generator of the drillers was indispensable to ensure the survival of the camp during the long wait for help, more than seven months away.

On a scientific level, the core samples from Vostok were not studied much at the time. The Leningrad Mining Institute was interested above all in drilling techniques, and their scientific colleagues from the Leningrad Arctic and

Antarctic Research Institute and from the Moscow Institute of Geography had very limited means. In 1975, Willi Dansgaard analyzed the content of oxygen 18 from surface samples, but a collaboration with the Russian teams did not occur for the analysis of the deep ice because, even though the measurements of that isotope were not precise, the Soviets had a mass spectrometer that enabled them to be carried out.

Thanks to personal contacts, in particular with Volodya Kotlyakov, director of the Moscow Institute of Geography, Claude Lorius initiated a collaboration among French and Soviet teams. The two teams had already established a strong friendship, so scientific arguments were easy to put forth. The laboratory in Grenoble was expert in the analysis of carbon dioxide, that of Orsay in beryllium 10. The analysis of deuterium done in Saclay appeared complementary and not competitive with those of oxygen 18. In 1982, Lorius went to Leningrad and returned with more than 700 vials of samples extracted between 0 and 1,400 meters, which were to be analyzed for deuterium. Jean Jouzel was delighted to pick them up at the Roissy airport but was disappointed a few months later when the measurements proved to be unusable. The vials in which they were preserved were of mediocre quality; this was not acceptable for isotopic analysis because the slightest evaporation changes the isotopic characteristics of a water sample.

It was a minor setback. At the end of 1984 Michel Creseveur, Claude Lorius, and Jean-Robert Petit landed at Vostok on an American C130. A few weeks later they had taken samples of more than two kilometers of ice cores and, above all, from completely virgin ice between 1,400 and 2,083 meters. The team was hoping that, using a simple calculation, done by assuming a constant accumulation, they would discover ice more than 100,000 years old, but as this accumulation decreased in a glacial period, the period covered was on the order of 150,000 years.[19] It was absolutely fabulous because we henceforth had access to a complete climatic cycle. The results, described in the next chapter, were astonishing. They were made public in 1987 in a series of three articles[20] in *Nature*, which guaranteed a very large promotion of the Vostok drilling, the results of which were called a "cornucopia." What pride for a European scientist to have Walter Sullivan, the famous scientific journalist of the *New York Times*, calling us to ask when he could come to France to meet us. There was just as much pride, less than five years later, in seeing that set of data in school textbooks.

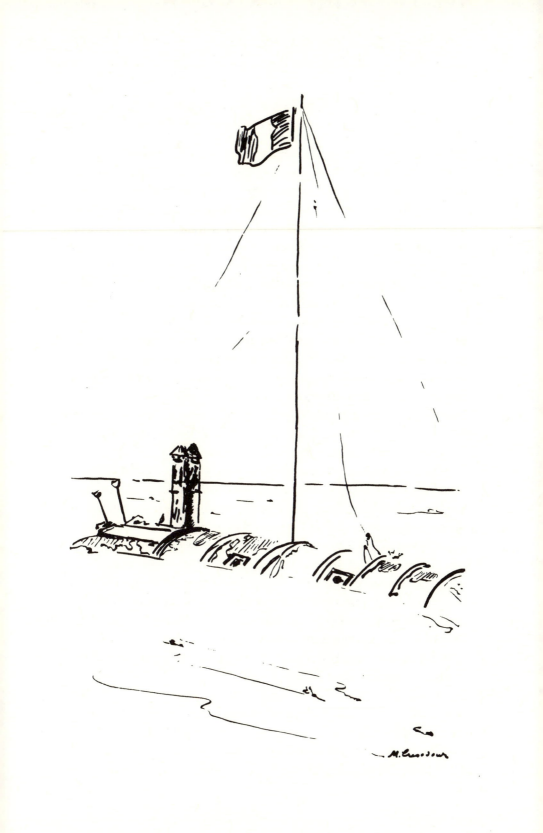

Europe and the United States: Two Drilling Operations in the Center of Greenland

The Vostok results received a lot of attention. A strong collaboration began between glaciologists involved in this project and the community of paleoceanographers because everyone saw the potential offered by the combination of data from marine sediments and those of Vostok. The ice core community also underscored the interest of drilling in the center of Greenland, the only zone that could cover a similar period. Willi Dansgaard had trouble convincing the NSF of the merit of such a project, especially since most American glaciologists were not active in the reconstruction of the climate from ice cores as they had had no access to deep ice since the Camp Century and Byrd drillings. He found an effective ally in Wally Broecker, a geochemist and oceanographer at Columbia University, who, like Hans Oeschger, was fascinated by the connection between the rapid variations discovered at Dye3 and the potential changes of the ocean current in the North Atlantic. Broecker began a campaign with the NSF and U.S. politicians for the United States to get involved in drilling in Greenland and with European scientists because he foresaw that such a project could only be launched within the framework of a large international collaboration.

Broecker in particular wanted the French to be involved in the operation. He called upon Jean Jouzel—who had worked with him and with NASA scientists in New York on aspects connected to the modeling of the variations of the contents of deuterium and oxygen 18 in precipitations—to ask Claude Lorius to join the project as well. Discussions took place in a New York restaurant on the occasion of Broecker's birthday. He fought to organize a gathering at the highest scientific level. This finally took place at the Concord Hotel in Boston at the end of January 1987. Dansgaard, Oeschger, and Jouzel, who was representing an absent Lorius, participated from the European side with, in addition to Broecker, four American scientists: Chet Langway, Harmon Craig (just as famous for his work on isotopes as Willi Dansgaard), John Imbrie (basking in the success of his astronomic theory), and Paul Mayewski (an American glaciologist ready to be involved in this project if it got under way).

Broecker opened the meeting by reiterating its objective: to launch a joint project to drill in the center of Greenland. Then Dansgaard spoke. For him, the objective was different: two core drillings, one European, the other American. There was a brief moment of astonishment, but the argument for two drillings was strong. The expected results were of such importance that it was indispensable to confirm them in parallel on a second drilling site. Discussion took off, and even if behind Dansgaard's proposal there was perhaps some hidden bitterness at not having been heard earlier by the Americans, the arguments for two sites were convincing. Those that were linked to the dynamics of the ice argued for an optimal distance of thirty kilometers between the two sites. From a logistical point of view, the establishment of two camps close to each other—a bit more than an hour apart by snowmobile—was reassuring. Finally, the cost of transports to central Greenland would be shared. The highest region of the Greenland plateau, which in its central part culminates at an altitude of 3,240 meters, was chosen: a very flattened dome, even though the glaciologists baptized the site Summit. The Europeans chose the highest point, the Americans a site 28 kilometers farther west at an altitude a few dozen meters lower. It was agreed that the two drillings would begin the same year and that the results would be shared in strict collaboration. Two projects, the American GISP2 and the GRIP (GReenland Ice core Project), were born.[21]

The success of Dôme C, the difficulties Dansgaard and Oeschger encountered in launching a project with the Americans, and the European Communities' official support for research on the environment and the climate all facilitated the emergence of European research in deep core drilling in polar regions in the 1980s. The idea of a three-way program (Denmark, France, and Switzerland) dedicated to the study of the last 1,000 years—the first 300 meters of the ice core at Summit Station—then took shape. It received very strong support from the European Communities and gave birth to the Eurocore project in 1989.

The European Science Foundation (ESF), whose headquarters are in Strasburg, was impressed by the objective of GRIP, which it decided to launch with national organizations responsible for ensuring it financially. Five other countries—Germany, United Kingdom, Belgium, Iceland, and Italy—joined the project, brought their scientific expertise, and, with the European Communities, helped finance the project. One of the unique aspects of GRIP was that it made laboratory space available to scientists that would enable them to study the many physical, electrical, and chemical properties of ice in the field. A scientific space nearly forty meters long and five meters wide was built for this in 1989 under a snow roof near the drilling trench. GRIP drilling started in June 1990. Various teams from Grenoble, Bern,

Rejkavik, Bremerhaven, and Cambridge were in charge of drilling alongside
the Danes. The core drill was permanent, and a system of three shifts working eight hours was adopted. Ice cores longer than two meters, a dozen per
day, on average, were brought to the surface. They were of excellent quality.
The core drill Istuk had been proven at Dye3. In spite of a few incidents the
drilling was carried out, as planned, over three summers: 710 meters in 1990,
2,320 meters in 1991, and 3,028.8 meters on July 12, 1992. Willi Dansgaard
arrived from Copenhagen to attend the extraction of the final core from a
drilling that marked the crowning of his career and to enjoy a wonderful
celebration. The Europeans were able to accomplish a great deal and even to
shorten the final season by a few weeks.

Our American colleagues had less luck. Their new core drill, made in
Alaska, was of impressive size. It enabled the extraction of ice cores six meters
long and of greater diameter (12 centimeters as compared to 10 with Istuk).
The drill was entirely satisfactory, but the cable proved problematic. After
three years of drilling, the season of 1992 had to be interrupted at a depth
of 2,200 meters. A new cable was brought to Summit in May 1993, and the
Americans reached the bottom in July at 3,054 meters. They beat the European depth record and, thanks to special tools, succeeded in extracting 50
centimeters of rock. GRIP and GISP2 were two logistical and technical successes by teams of very competent and cohesive drillers. The scientific results
were remarkable for the information they provided about the rapid changes
in climate, but they remained limited to the last 100,000 years because beyond that the flow of the layers of ice was disturbed by the proximity of the
rock base. These disturbances were revealed thanks to access to two neighboring ice cores. In retrospect, the choice made in Boston was fully justified!

Europe Turns to Antarctica

Like Willi Dansgaard in Greenland, Claude Lorius pursued a specific objective in Antarctica: to reach the bedrock at Dôme C. That site was a priori
even more favorable than that of Vostok. Flow models indicated that the
deep ice could be more than 500,000 years old. A long-term effort was
started after the first drilling at 900 meters. The thermal drill option was adopted. In 1987 that drilling system was tested with success at site D47 in the
coastal regions of Adélie Land. But that operation also revealed the complex-

ity of manipulating the drill, which also had the disadvantage of being cumbersome and thus difficult to transport to the center of Antarctica. Logistical matters were, however, of primary concern for the French. For the summer season to be long enough at Dôme C, it was necessary for the logistical, technical, and scientific personnel to arrive very early at Dumont d'Urville and leave from it as late as possible. Then plans for a permanent base on the site of Dôme C, base Dôme Concordia, were conceived, the interest in which went well beyond the community involved in the study of the atmosphere and the climate since it was meant to be open to researchers in the realms of geology, the upper atmosphere, astronomy, biology, and medicine.

For these reasons, the French scientists who had projects on the Antarctic continent were generally in favor of the construction of the landing strip on Adélie Land by the EPF under the direction of the Territoires des Terres Australes et Antarctique Françaises (TAAF). Except for those, such as biologists, who were concerned about the impact of the airstrip on the local ecology, it was very disappointing to see it soon damaged and then abandoned. Salvation came from Italy, which had decided to abandon research in nuclear technology, resulting in a reorientation within the Ente per le Nuove Technologie, l'Energia e l'Ambiante (ENEA). Some of its teams were redirected toward Antarctic research. An Italian base was built at Terra Nova Bay, located at the same distance from Dôme C as Dumont d'Urville. But that base had the advantage of being accessible at the beginning of the season by large transport planes, which, leaving from Christchurch in New Zealand, landed on a strip prepared on the sea ice. The Italian scientists, led by the very dynamic Mario Zuchelli, were natural partners for their French colleagues. An agreement was signed between the ENEA and the new Institut Français de Recherche et de Technologies Polaires (IFRTP), which took over from the EPF in 1992. The Dôme Concordia project and that of deep core drilling at that site became a Franco-Italian venture.

The drilling project quickly took on a European dimension, primarily for scientific reasons. The success of GRIP had created very strong bonds, in particular for those who were fortunate enough to spend long months together in the field. Scientists and drillers from various European labs were very motivated by an ambitious project in Antarctica. It was also the only path that enabled them to obtain European subsidies, especially since other countries such as Germany and England, who envisioned drilling in other regions of

Antarctica, also fell under the European umbrella. Claude Lorius and David Drewry, director of the British Antarctic Survey, and Heinz Miller from the Alfred Wegener Institute in Bremehaven (Germany) played a major role in merging these various initiatives in a common European project. This was done thanks to a series of meetings between all European scientists interested in such a project. One ended with the idea of two complementary sites in East Antarctica, that of Dôme C and another to be located in the Atlantic sector, which was then completely unexplored, so as to enable an optimal comparison with the records in Greenland. Many reconnaissance campaigns were necessary to identify an appropriate drilling site in that vast region of Dronning Maud Land (DML at Kohnen Station) and it was thus natural to begin the project, which was baptized European Project for Ice Coring in Antarctica (EPICA) (of which Jean Jouzel became the coordinator), with drilling at Dôme C.

In 1993, a high-level conference was organized at a more political level in Bremen (Germany) in order to identify European "Great Challenges" in the field of environmental research. This label was recognized as in association with EPICA, which was very helpful for the proposition established under the aegis of the ESF to get financial support from Europe. Ten countries were involved: Germany, Belgium, Denmark, Italy, France, Norway, The Netherlands, England, Sweden, and Switzerland. The electromechanical drill was chosen over the thermal one that we had used, and an EPICA drill, which was based on the Danish Istuk model, was developed in collaboration with Danish, French, Italian, and Swiss teams. The 1995–96 season was devoted to developing the necessary measurements to select the precise site, approximately fifty kilometers from the former Dôme C drilling. The following season material was transported overland between Dumont d'Urville and Dôme C in difficult conditions due to crevasses in coastal areas and to weather conditions by teams from IFRTP under the direction of Patrice Godon. Other personnel had been brought by Twin Otter from Terra Nova Bay. That season culminated in the establishment of the camp and in the creation and the tubing of a "fore-hole" of 130 meters needed to start deep core drilling. In the wake of the success of GRIP and in anticipation of similar success at Dôme C, a "scientific tent" was built and outfitted for the scientists. In November 1997 everything was ready so that the new Dôme C deep core drilling could begin.

Vostok: More than 3,600 Meters of Ice

Still in Antarctica, let's go back fifteen years. The fire that destroyed the generator of Vostok Station on April 12, 1982, could have had tragic consequences, but it did not alter the enthusiasm of the Soviet drillers. In 1984 they undertook a fourth core drilling from the surface. At the same time they attempted a deviation on the one that they had had to stop. The maneuver worked, but a few months later the drill was blocked and that core drilling had to be abandoned in November 1985. All hopes rested on the new hole. Operations continued relatively slowly but without problems until February 1990. But once again bad luck struck, and the drill was blocked at a depth of 2,546 meters. Those 500 additional meters offered an extension of some tens of thousands of years of climatic records. Interesting, of course, but above all frustrating because there was still a kilometer of ice to be discovered, a simple calculation of which showed that it covered two or even three additional climatic cycles.

The project might well have been abandoned, but nothing of the kind occurred. The drillers' enthusiasm never waned. Also, the collaboration was enlarged to include American teams, which meant the continuation of the indispensable logistical support of the NSF. This support was as unwavering as that offered by the IFRTP, led by Roger Gendrin, on the French side. While our Russian colleagues from Saint Petersburg and Moscow were experiencing very difficult times in the years following the end of communism in the USSR, operations at Vostok were never threatened. Logisticians, drillers, and scientists got together each year, alternatively in Russia, France, and the United States. The decision was quickly made by this international consortium to start again from the surface. This was done at the end of 1990. The drilling was ultimately successful, but it took eight years: the station had to be closed twice during the winter because it had not been restocked with fuel, largely a result of the financial difficulties of our Russian colleagues. The project's success was due in large part to the expertise of the drillers from the Saint Petersburg Mining Institute, under the direction of Nicolay Vasiliev, who efficiently alternated thermal and electromechanical drilling. It was also due to the heavy involvement, alongside that team, of two scientists who adopted the project: the Russian Volodya Lipenkov and the French Jean-Robert Petit.

The core samples were shared among the three countries, but before they were transported a thin slice was cut along each of them. Once melted, the water was stored in flasks intended for the Saclay laboratory where those precious samples were analyzed as quickly as possible.

A depth of 2,755 meters was reached in January 1994; the core samples then covered two climatic cycles. Then 3,350 meters was reached in January 1996—600 additional meters, which doubled the period with an estimated age of 420,000 years at 3,310 meters. We were disappointed because data suggested that below that level the layers of ice, as in the center of Greenland, were disturbed. Core drilling continued, but we decided that it should be stopped a bit beyond 3,600 meters because radar measurements taken from the surface had revealed the existence of a huge subglacial lake the size of Lake Ontario or Corsica, with a depth of 600 meters, directly below Vostok Station. It was obviously essential to avoid contaminating that water, which could have been more than a million years old. Core drilling was stopped in December 1998 at around 120 meters above Lake Vostok and close to thirty years after the operations began.

Since then our Russian colleagues have been extremely motivated by sampling liquid water from Lake Vostok, an undertaking that raised a great deal of concern among the international glaciological community because of the high risk of polluting the lake water. The Russian drillers claimed that such contamination would be avoided—or at least very limited—because of the overpressure at the lake surface when it would be reached. They convinced their authorities and were given permission to proceed. Under their initiative, drilling of the core resumed in 2009 by a deviation of the hole reaching a depth of 3,720 meters in December 2011. Nonstop drilling operations began on January 2, 2012, and continued until February 6 under the leadership of the Russian chief driller, Nikolay Vasiliev. The last flight ending the summer season at Vostok was scheduled for February 6. We summarize here a report of the last day of operation by our Russian colleague Volodya Lipenkov, who was part of this field party: "the drill hit the surface of Lake Vostok in the morning of that same day at a depth of 3769.3 meters. Due to the high water pressure at the surface of the lake, liquid water surged up the borehole, and one minute after the penetration the kerosene used as drilling fluid began outflow through the top of the hole. This shows that we had reached and broached the main water body of the lake without contaminating it. This outflow persisted for about five minutes and about 1.5 cubic meters of drilling fluid was forced out the top of the casing. According to our estimate the water rose to about 600 meters above the lake surface. The last drilled core was mostly frozen and was sampled for biological and isotopic analyses just a few hours before we left the station with the last flight." Lipenkov added, "Such was the end of the 5G drilling but hopefully not the end of the Vostok project." Next season (2012–13), the Russian team plans to come back to the station to start coring the lake water frozen in the hole.

Other Core Drilling in Antarctica

The decade of the 1990s was full of other coring activity in Antarctica. The Japanese were initiated in coring methods through contact with Europeans during the GRIP coring. Evenings under their tent, which was set up a few hundred meters from the main dome, were congenial and warm. Their participation in the GRIP campaigns was fruitful since they built their own drill inspired by Istuk, which quickly became operational. The project initiated by Okitsugu Watanabe was ambitious: to reach the bedrock at Dôme Fuji, an ideal site on the other side of East Antarctica from Dôme C. As at Vostok, it was decided to drill throughout the year in spite of winter temperatures below −70°C. In two years, 1995 and 1996, the Japanese drillers were at 2,503 meters. The level of the drilling liquid was then much lower than what it should have been. Perhaps that was the reason the drill became blocked; despite every effort it was impossible to bring it up. There was success, however, because the deepest ice was around 330,000 years old, and three complete climatic cycles were recorded. But there were also many regrets because we now know that the extraction of the 500 remaining meters would have allowed the record held by Vostok to be broken. The Japanese did everything they could to resume drilling from the surface, and that new operation was crowned with success at the beginning of 2006 when they obtained a 3,035-meter-deep core covering more than 700,000 thousand years.[22]

Other teams chose more modest projects, aiming to drill on one of the small domes that exist in the coastal regions. There is a good deal of snow precipitation there, which helps in dating the ice because the years can be counted. There is also an advantage for the interpretation of records contained in the air bubbles because, due to this high accumulation, the uncertainty about the difference in age between these bubbles and the ice that contains them is much less than in regions on the Antarctic plateau with little accumulation. The drillings on small domes, whose depth generally did not go beyond 1,500 meters, were in fact very well adapted for the study of the relatively recent climate, say the last 20,000 years, a period that covers the Last Glacial Maximum, the deglaciation, and the Holocene. Some of those ice cores cover all of the last glacial/interglacial period and even beyond.

Above all, these small domes are located in regions that are relatively accessible and thus required limited logistical support. For example, the Australians did not have the necessary logistical support to mount an operation far from their base, Casey, but they were fortunate to have a little ice dome, Law Dome, near that station. A single summer season (1996–97) was necessary for a 1,200 meter long core to be extracted, which was particularly interesting. More recently, drillers from Grenoble, with the support of the IPEV under the direction of Gérard Jugie (Institut français polaire Paul-Émile-Victor, the new name of the IFRTP since the beginning of 2002), developed a new core drill, light and easily transportable, used in collaboration with the British Antarctic Survey (Cambridge) to carry out an operation on Berkner Island; the drilling, 948 meters deep, reached the bedrock in January 2005. The same drill was then used to obtain, within the framework of an international project coordinated by Italy, a core of about 1,500 meters at Talos Dome, a site close to the Terra Nova Bay Station. The goal was reached at the end of 2007.[23] At the beginning of 2008, there was also success on James Ross Island, to the east of the Antarctic Peninsula, whose glacial ice sheet was drilled down to the bedrock.

It is somewhat surprising that the logistical means the NSF could devote to glacial drilling were, in fact, rather limited. That organization, unlike the IPEV, did not have the means to transport materials overland, and the transport of all supplies in Antarctica was dependent on the use of large C130 transport planes from McMurdo Base. But those aerial means were primarily devoted to transporting the material necessary for the reconstruction of the permanent base on the South Pole and for an experiment aiming to count the very rare solar neutrinos using the ice as a detector. The glaciologists were therefore forced to concentrate on sites that were easily accessible from McMurdo. Two core drillings were thus carried out at Taylor Dome between

1991 and 1994 and at Siple Dome, where the bedrock was reached in January 1999. Those drillings were primarily interesting in terms of the study of the last 100,000 years. The Americans then decided to launch a large-scale operation in West Antarctica, a region particularly interesting in which to check the assumption put forward by some scientists who had feared that it had been unstable in the past. Deep core drilling in the central regions of this part of Antarctica would enable a better evaluation of this hypothesis. But our colleagues needed a lot of patience because, at the beginning of the 2000s, logistical support still was largely devoted to activities in the South Pole. Drilling operations of the West Antarctic Ice Sheet (WAIS) ice core began in 2008, 160 kilometers from the location of the Byrd ice core. A depth of 2,561 meters was reached in January 2010 and the drilling was completed on December 31, 2011. The final core was collected at a depth of 3,405 meters: the WAIS team could not have asked for a better reason to celebrate the New Year. Due to the high accumulation at this site,[24] the core does not extend beyond the last 62,000 years, but it will allow scientists to obtain records of very high accumulation.

The Glaciers of the Andes and the Himalaya

We are indebted to an American team, that of Lonnie Thompson of Ohio State University, for having been the first to launch extensive ice drilling projects in other regions. He was the first to believe in the scientific value of cores extracted from tropical glaciers, which many of us doubted, and to drill above 6,000 meters in altitude in the Andes and the Himalaya. Of course, just as in the coastal zones of Antarctica, the scale of time was limited to tens of thousands of years, but the Andean glaciers are excellent archives of the past rhythm of the El Niños and the Himalayan glaciers of that of monsoons. In addition, some of those glaciers are shrinking in volume at a dangerous rate and quite simply risk disappearing under the effect of global warming. It is worth pursuing this area of study, but there are difficulties in drilling even if the thickness of the ice to be cored never goes beyond 200 meters and in retrograding the ice in the laboratories. Light drills that run on solar energy have been developed, but everything must be transported by foot from the last accessible base. And once the cores are extracted, they have to be brought back down as quickly as possible to avoid melting. Lonnie Thompson overcame all those difficulties, and we are indebted to him for an entire series of ice cores from those tropical glaciers.

Other teams have followed in his footsteps: that of Paul Mayewski in the United States and a Franco-Swiss consortium led by glaciologists at the Institut de Recherche et Développement (IRD). This consortium, in which on the French side the Laboratoire des Sciences du Climat et de l'Environnement (LSCE) Saclay and the LGGE Grenoble collaborate, has succeeded in extracting a core in Chimborazo, just south of the equator at more than 6,000 meters in altitude. But the most amazing project is that of the Chinese: to extract an ice core from the Muztagh Ata, on the roof of the world, at more than 7,500 meters in altitude. Closer to home, in the Alps, there is little hope of going back to past climates from glaciers because they are generally places of surface fusion and percolation, phenomena that blur the signals recorded by the snow. On the other hand, these temperate glaciers provide an excellent means for reconstructing the increase in pollution connected to the advent of the industrial era in Western Europe.

A Return to Greenland

There were other drilling projects in the 1990s. The Danes were very disappointed that the core samples from the center of Greenland, GRIP and GISP2, did not reliably allow them to go back further than 100,000 years. There was still hope, however, because it was possible that the mixture of layers observed at the base of these cores was linked to the fact that the subglacial relief was very tortured in that zone. It would therefore suffice to locate a region with a flatter relief to be able to go back further in time. In 1994, reconnaissance missions were organized in the northern part of Greenland. A zone located 200 kilometers north of GRIP appeared a priori favorable. The team from Copenhagen assumed the leadership of a new coring project baptized North GRIP, in which we participated alongside German, American, Belgian, Japanese, Scandinavian, and Swiss teams. The camp was established in 1995, and core drilling began successfully in 1996. Everything was going well, but the following season the drill was blocked at a bit more than 1,400 meters.

Nothing could be done; neither the efforts undertaken that year nor those undertaken the following season could bring the Istuk drill back to the surface. The team decided in 1999 to start again from the surface. In two seasons a depth of 2,931 meters was reached, but the drilling was subsequently greatly slowed down. Indeed, the geothermal flow was higher than anticipated, and the temperature of the ice was closer to the point of fusion than was predicted by flow models. To drill into that ice, which is called "warm," required enormous precautions, and procedures had to be put into place to ensure the drill did not get stuck. Three seasons were needed to reach the bedrock; this occurred in 2003. More specifically, the drilling ended in a subglacial "river" present under 3,085 meters of ice, with as a consequence the rise of 45 meters of that water, subjected to very strong pressure, before it froze.

The presence of liquid water at the base of the glacier had a beneficial consequence because the relatively great fusion at the base of the North GRIP drilling contributes to making the deep layers disappear. The thinning of the layers of ice within the ice sheet was reduced and, with that, the risk of a perturbation due to the ice flow like those that affected the last 300 meters of the GRIP and GISP2 drilling. The age of the deepest ice was estimated at 123,000 years, probably the warmest time of the last interglacial period.[25] The drilling thus did not go back as far as had initially been hoped for, but the disadvantage was largely compensated by the absence of perturbations linked to the ice flow. In a rather unanticipated way, we thus went from obtaining deep cores of which approximately the deepest 10% were practically unusable from a climatic point of view (in both Antarctica and Greenland) to a completely unperturbed record.

Empowered by that success, the Copenhagen team, with their usual partners, initiated a new drilling project at a site located farther north, between North GRIP and Camp Century. New teams (Canadian, Chinese, and Korean) joined the project, which began in 2007 with the establishment of the camp and ended in 2010 with a core drilling that reached the bedrock on July 27 at a depth of 2537.36 meters. The hope was to cover the entire previous interglacial period, but unlike for North GRIP the bottom part is perturbed by ice flow below about 2.2 kilometers. However, available data show that a correct time sequence can be reconstructed back 129,000 thousand years, extending the North GRIP record over a large part of the last interglacial period.[26]

The European EPICA Drilling: A Double Success beyond All Hopes

The years 1997 and 1998 were extremely trying for all the drilling teams with the loss of the drills both at Dôme Fuji and at North GRIP. And the EPICA Dôme C project wasn't spared either. The first season (1997–98) was difficult because it was the true baptism by fire for the EPICA drill. Drilling resumed the following year and was satisfactory, but after a few weeks the drill was stuck at a depth of 780 meters. Despite all the attempts made by the drillers brought in by Laurent Augustin from Grenoble, it was impossible to recover the drill and to resume operations. It was a huge disappointment for everyone because they had to begin again from the surface, and that solution didn't appeal to the drillers, the scientists, or the institutes and organizations that had helped finance EPICA. In spite of everything, the drilling operation was resumed during the 2000–2001 season, and it was henceforth proven that the decision to resume drilling after that failure was a good one.

After the 1999–2000 season, which was devoted to putting new tubing into place and testing the changes made to the drill, the success of the two following campaigns exceeded all expectations. A depth of 1,459 meters was reached at the end of January 2001, and the final following season saw a doubling of the depth at 2,871 meters. A comparison of the records of electrical conductivity indicated an age older than that of 420,000 attained at that site. The objective of obtaining 500,000-year-old ice had, in fact, already been surpassed. However, the increase in temperature, due to the influence of the geothermal flux as one went farther into the ice sheet, made the drilling conditions increasingly difficult. In spite of these difficulties, more than 300 meters of ice were extracted during the 2002–2003 season, and drilling was finally stopped at a depth of 3,201 meters. After stopping for a season, the EPICA Dôme C drilling reached the bedrock at 3,260 meters in January 2005. Ice drilling at Vostok kept the world record of depth, but even if the measurements taken from the deepest ice indicated that the last 60 meters had been perturbed, the Dôme C drilling was enormously satisfying, allowing us to go beyond 800,000 years.[27] For a time, the Japanese hoped that the second drilling at Dôme Fuji would contain ice that was one million years old, but that estimation was lowered and the age close to the bedrock was on the order of 700,000 years. Dôme C thus holds the record for the longest climatic sequence, a record that will likely remain unbroken for several years.

The second EPICA (Kohnen) drilling was undertaken in 2001–2, the season during which a depth of 450 meters was reached; it continued without difficulties and reached a depth of 2,565 meters in 2004. The ice there was around 200,000 years old. That drilling, which should have enabled us to obtain extremely detailed records over an entire climatic cycle, reached the bedrock during the 2005–6 season, at a depth of 2,760 meters. This double success opened the door to an entire series of articles describing our climate and our environment during the last 800,000 years and to a group of new results bearing on the climatic interactions between the Northern and the Southern hemispheres. This wealth of results earned EPICA the honor of being one of three projects to receive the prestigious Descartes Prize for Science given by the European Union in 2008.

It has now been nearly fifty years since the first glacial deep core drilling began at Camp Century. Fifty years during which the community of glaciologists has not been spared failure and disappointment but which, above all, have been full of success, to which collaborating drillers, logisticians, and scientists have all contributed. This has been a period marked, perhaps more than for other scientific research fields, by the increasingly international nature of the programs. Thus, on the French side, in addition to the collaborations with the Russians, then the United States in Vostok, with a European consortium for GRIP and EPICA, within the framework of a multinational program for North GRIP and NEEM, there have been other partnerships such as those we have established with Australia in the drilling at the Law Dome and with Japan in that of Dôme Fuji. We should also mention the first drilling achieved on James Ross Island, off the Antarctic Peninsula, in collaboration with Argentina, as well as the projects with the Chinese. This diversity has enabled French teams to be involved in almost all the great advances that have been made in the study of ice core samples, and in particular in the two scientific discoveries that we discuss later in the book: the existence of a relationship between greenhouse gases and the climate in the past (chapter 11) and that of rapid climate changes (chapter 9).

CHAPTER 7

Vostok

THE CORNUCOPIA

The term *cornucopia*, attributed by the journal *Nature* to the Antarctic Vostok ice core drilling, deserves explanation. In the mid-1980s, our knowledge of the great glacial/interglacial cycles that marked the Quaternary essentially rested on the study of marine sediments. Thanks to them the astronomical theory, which stipulates the existence of a connection between the variations in insolation linked to the slow evolution of the Earth's orbit and those great climatic cycles, had been very widely accepted. In the wake of the article published in *Science* in 1976 by Jim Hays, John Imbrie, and Nick Shackleton, who established the connection, paleoceanographers put an ambitious program into motion: the Climate/Long Range Investigation Mappings and Predictions Project (CLIMAP) aimed to map the climatic conditions that ruled over all the oceans in the Last Glacial Maximum.[1] The data, published during the 1970s and the beginning of the 1980s, fully reveal how marine sediments have aided in our knowledge of past climates.

Ice core drilling and science in Greenland and Antarctica were still in their infancy at this time. Thanks to high accumulation, ice core drilling at Byrd offered a very great thickness of ice for a detailed study of the last deglaciation, which began around 20,000 years ago, and of the warm period in which we have been living for more than 11,000 years. On the other hand, the ice of the last glacial period proved difficult to date beyond 60,000 years. The situation was rather similar for the two core samplings from Greenland that were available then, those of Camp Century and Dye3, for a simple reason: the oldest ice corresponded to the deep part of the core, the final 150 to 200 meters near the bedrock. The flow of the ice becomes increasingly complex the closer one gets to it, to the point that it is impossible to develop a flow model sufficiently reliable to establish a chronology.

The first ice core drilling at Dôme C escaped that problem: the maximum depth reached, 903 meters, was located more than two kilometers above the base. Calculating the flow was simple, especially since the site is located on a dome. The flipside was that this core sample from Dôme C did not go back very far in time. Even if the age of 30,000 years that we had initially attributed to it later proved to be too young (it has henceforth been established that the core retrieved in 1978 covered the last 45,000 years), we were still far from the beginning of the last glacial period, around 110,000 years ago. Thus neither the drilling at Camp Century or Byrd, done during the 1960s, nor those of Dôme C in 1978 or Dye3 in 1981 could rival the marine sediments available at the time. The oldest ice was too young for the results it provided to revolutionize our knowledge then of the succession of glacial and interglacial periods that extended over hundreds of thousands of years.

However, those ice cores did enable us to document the climate of Greenland and Antarctica in an extremely detailed way. The isotopic records obtained in Greenland put the researchers of Copenhagen and Bern on the path of rapid variations. Other parameters such as those corresponding to the fall of desert dust and of sea salt were also recorded in various ice cores; they bore witness to a very active atmospheric circulation during the Last Glacial Maximum. What is more, the core drilling at Dôme C was the occasion of many firsts. Among them we have already cited the possibility demonstrated by the team from Orsay of analyzing beryllium 10, an isotope of cosmogenic origin produced through the action of cosmic radiation on upper layers of the atmosphere, in the polar ice. There was also the isotopic analysis of the oxygen of air bubbles extracted from ice using a method perfected by our American colleague Michael Bender during a stay at the Centre des faibles radioactivités in Gif-sur-Yvette[2] and the development of a joint profile of concentrations of deuterium and oxygen 18 in the ice that was carried out at Saclay.[3] The most important result was that obtained by Robert Delmas and his colleagues at the LGGE: in the last glacial period, concentrations of carbon dioxide were 30% weaker than those that existed just before the industrial revolution.[4] With some distance, all of these approaches proved important and were at the origin of much later work, but our community of glaciologists was a bit behind because none of the ice core drilling went beyond the last glacial period.

A Complete Glacial-Interglacial Cycle

The situation changed dramatically between 1985 and 1990 for two reasons. Thanks to the ice core drilling at Vostok, French researchers had access to a core sample that had gone beyond a depth of two kilometers. It was enough to obtain the ice formed during the preceding glacial period, 150,000 years ago. Despite everything, that was far enough from the bedrock for a viable chronology to be established. And then those new methods put into place at the ice core drilling at Dôme C could be applied on that core sample as well as new approaches, especially since the LGGE in particular, thanks to Jérôme Chappellaz, quickly mastered the analysis of the composition of methane that was initially developed in Bern.[5] The unique possibility that the Antarctic ice provided in the reconstruction of the history of the two principal greenhouse gases whose composition was affected by human activity meant the work carried out on the Vostok ice would have many repercussions well beyond the relatively small circle of scientists who were interested in the history of our climate.

The reconstruction of climatic variations based on the isotopic analysis of the ice drew attention in 1985. At the coldest point of the last glacial period the temperature was close to −65°C (Figure 7.1); it was thus close to 10°C colder than it is currently (in Vostok the mean annual temperature is −55°C). It is interesting to note that the cold periods have a periodicity of close to 40,000 years, a periodicity characteristic of obliquity, a parameter of the Earth's orbit that determines the annual variations of insolation. Further, these cold periods occur at the moment when local insolation is the weakest. Thus part of the mechanisms put into play are different than those concerned in Milankovitch's theory, based on the variations of summer insolation in the Northern Hemisphere governed by the change in precession. But from this record of temperature at Vostok we had a clear indication of a connection between insolation and climate. The last interglacial period, around 130,000 years ago, was warmer at Vostok than it is today (by 3–5°C), and it probably lasted longer than in other regions of the planet. Furthermore, the variation in temperature at Vostok presented similarities with that of the sea level, which confers a global character, at the very least from a qualitative point of view: when the sea level is low due to a large amount of ice on the ice sheets of the Northern Hemisphere, it is cold in Antarctica, and vice versa.

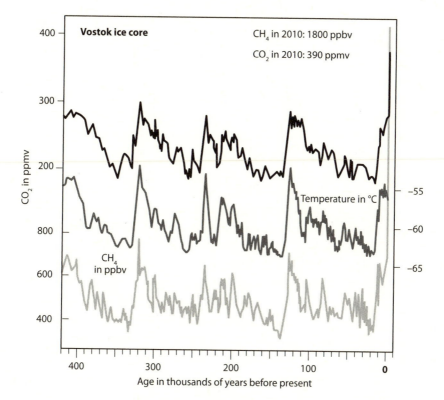

Figure 7.1. Vostok ice core. The middle curve indicates the variation in temperature at Vostok throughout the last 420,000 years. The upper and lower curves represent concentrations of CO_2 and CH_4 over the same period, indicating the recent spike of these two greenhouse gases.

Climate and Greenhouse Effect Go Hand in Hand

Was there a high concentration of CO_2 during that warm period 130,000 years ago? Was the cold period that preceded it, the next-to-last glacial period, characterized by concentrations comparable to those of 20,000 years ago? The response so anticipated by the scientific community was positive and unambiguous (Figure 7.1). But because data bear on an entire climatic cycle, the ice of Vostok provided a much stronger message: the records obtained along the more than two-kilometer ice core indicated that throughout the last 150,000 years the concentrations of carbon dioxide were, in general, faithfully correlated with the variations in temperature deduced from an isotopic analysis of that ice; the colder it was, the weaker the concentrations, and vice versa. On October 1, 1987, that which would be unanimously greeted as a discovery made the cover of *Nature*, which uncharacteristically devoted three articles to the Vostok ice core in that issue. The first presented and discussed the detailed profile of the variations in temperature at Vostok; the second focused on the record of carbon dioxide concentrations; and the third analyzed the connection between those two parameters, which vary in a parallel way, and orbital forcing. These three articles, already cited in previous chapters' notes, were accompanied by two commentaries, one by Eric T. Sundquist, a specialist in carbon cycles, the other by the journal's editor, Philip Campbell, to whom the ice core drilling of Vostok owes the name "cornucopia." The media, recognizing the importance of the relationship between carbon dioxide and past climate change, seized the results that, three years later, were reinforced with data relative to the concentration of methane; it, too, appeared strongly correlated to the temperature in Antarctica, with values twice as great in the warm period as in a glacial period. These curves of Vostok (such as those depicted in figure 7.1 showing the relationship between greenhouse gases and climate) traveled around the world through political as well as scientific circles.

We have mentioned the confirmation provided by the temperature curve at Vostok for the existence of a link between insolation and climate, but the analysis of the air extracted from the ice also very strongly supports the idea that the variations in greenhouse effect have played an important climatic role in the past. An awareness of variations in concentrations of carbon dioxide and methane appears to fill two of the main gaps in the astronomical theory of paleoclimates, which faces a dual difficulty. How can we explain the 100,000-year cycle that dominates climatic records but is present in the astronomic data only through the eccentricity whose influence is very weak? To what is owed the relative synchronism of the climatic events between the two hemispheres when variations in insolation are not synchronous? It is in fact toward the idea of a climate whose astronomical parameters would be the metronome—and where the greenhouse gases, and especially CO_2, would play the role of amplifiers of changes in insolation—that we have been led by Antarctic records. However, the explanation of these natural variations in concentrations of carbon dioxide and methane is only partial at this point. It has still not been settled convincingly, but we know that, in each case, the variations of these two greenhouse gases put into play interactions between physical parameters of the climate and geochemical cycles linked to the living world. For carbon dioxide, these variations certainly imply the circulation of the oceans and their productivity and, to a certain degree, interactions with the continental biosphere. On the contrary, variations in methane essentially come from the impact of the climate on its earthly sources, including swamp zones and perhaps the frozen ground of high northern latitudes.

Whatever the causes might be, the natural variations in concentrations of these two compounds bring about an increase in the greenhouse effect by a bit more than 2 Wm^{-2}, thus the equivalent of that linked to human activity during the last two hundred years. If the climatic system was exempt from feedbacks, that direct radiative effect would produce an increase in temperature of only 0.6°C. But sea ice, water vapor, and clouds in a glacial period—and as is the case today and as it will be in the future—modified the response of this system. Indeed, the sensitivity of the climate to these so-called rapid feedbacks appears to depend weakly on the climatic period under consideration, and we proposed to estimate this parameter from Vostok and other data of the past. This approach does not require that the complexity of the mechanisms driving large climatic changes be completely deciphered or that a definitive response to the "chicken or egg" question—which is the cause, which is the effect—be answered, which is so often raised on the subject of respective variations of the climate and of greenhouse gases during the past. When we are looking only at the sensitivity of the climate, it is enough that the various forcings that operate on the scale of large observed climate changes can be correctly estimated; then we evaluate the role of the various factors that can account for the succession of the glacial and interglacial ages.

Although a simplistic approach, we used a means to statistically compare the temperature profile of Vostok with different forcings able to explain it: the astronomical contribution represented by the variation in volume of ice accumulated in the Northern Hemisphere and by that of local insolation, radiative forcing linked to the variations in CO_2, modifications of the radiative budget resulting from changing the content of atmospheric aerosols (dust and sulfates). This statistical analysis confirms the visual impression: a contribution linked to the astronomical forcing (via the variations in volume in the ice in the Northern Hemisphere) and greenhouse effect each explain

40 to 50% of the variability in Vostok temperature; the influence of other possible causes remains marginal. Thus an increase in the content of greenhouse gases would explain up to 2°C of the 4–5°C of average warming of the Earth during deglaciation, indicating that the direct radiative forcing was increased, through feedbacks linked to sea ice, water vapor, and clouds, by a factor close to 3 (~2/.6). In the case of a doubling of the content of carbon dioxide compared to its approximate current value, the forcing would be double that observed since the Last Glacial Maximum (4 Wm^{-2}), and we have deduced from the data of the past that the warming would then be 3–4°C. This approach has limits, both on the level of statistical method put into place and in terms of estimating accurately the variation in mean temperature of the planet in the past. However, it illustrates the key result provided to various degrees by the climatic models whose most recent results concur with the deductions we published in 1990[6] from data issued from polar ice: these are mechanisms of amplification vis-à-vis the radiative forcing connected to the anthropogenic greenhouse effect, which should operate during the coming decades. Furthermore, the results obtained at Vostok have undeniably played a role in the awareness of the link (discussed in the last part of this work) between the evolution of our climate and the increase in the greenhouse effect related to human activity.

One, two, three, and then four climatic cycles were finally reached in 1999. The records we have obtained, along with our Russian and American colleagues, thus cover 420,000 years. The theory put forth a dozen years earlier was fully confirmed: the climate of Antarctica and greenhouse gases go hand in hand throughout the period characterized by the succession of four very marked glacial/interglacial periods.[7] This extension of the records highlighted the role of human activity during the recent past. As observed along the Vostok core, never in 420,000 years have the quantities of carbon dioxide and methane present in the atmosphere been as high as they are today (Figure 7.1); never have they risen so rapidly.

Thanks to the comparison of records obtained in the Vostok ice and those provided by the analysis of the ocean sediments, we began to be able to better decipher the complexity of the mechanisms of large climate changes. Thus we henceforth had a better view of the sequence of events as it was repeated during the transitions between a glacial and interglacial period. Such a transition corresponds to a global warming on the order of 5°C on average, accom-

panied by the melting of most of the ice accumulated on the continents of the Northern Hemisphere, a melting that provoked a rise in sea level by about a hundred meters. Somewhat surprisingly, things may have begun to move in the southern regions: in Vostok the first signs of warming were observed—probably linked to subtle changes in insolation—but, although the relative timing between temperature and CO_2 at Vostok is difficult to estimate, the increase in the concentration of carbon dioxide possibly began a few centuries or a thousand years later. Whereas the temperature in Antarctica and the amount of carbon dioxide (and methane) increased, it was only after a few thousand years that the ice sheets of the Northern Hemisphere began to melt so extensively. This process ended several thousand years later. Was it the local insolation, that which prevailed in Antarctica, that gave the signal to begin? Or did that result indirectly from changes in insolation in the Northern Hemisphere, as proposed by the classic astronomical theory? This question has not been definitively answered.

Everything leads us to believe that these changes in insolation modified the conditions prevailing in the Southern Ocean, a large reservoir of carbon, which was then translated into the observed increase in the concentration of carbon dioxide. Once in play, it participated in the warming by increasing, through the radiative forcing associated with it, that which was initiated by the change in insolation (Figure 12.5). The sequence of events was such that the increase of the greenhouse effect then participated fully in the melting of ice sheets in the Northern Hemisphere, itself amplified by the feedback linked to the change of albedo (the gradual disappearance of ice sheets diminishes the reflective surfaces and accelerates their melting). To the question "Which came first, the chicken or the egg?" the answer is not so simple. In our opinion, it was indeed radiative forcing that was the initial cause of deglaciation, but once the greenhouse effect began to increase, it in turn became a true agent in deglaciation—to the point of being half responsible for the mean increase in the temperature that accompanied these major climatic transitions.

Much More Information

Much more information has been revealed by the ice of Vostok, all of which are "firsts," since, up until 2002, that glacial ice core was the only one that allowed us to obtain reliable records beyond 100,000 years. We will mention just a few of the results that have undeniably advanced our knowledge of the conditions that prevailed since 420,000 years ago. We have already mentioned the spike in production of beryllium 10 that occurred around 40,000 years ago, discovered by Grant Raisbeck and Françoise Yiou. Jean-Robert Petit has shown that the fallout of desert dust—which depends on the extension of arid zones and the intensity of atmospheric circulation—was systematically different between cold and warm periods with very high levels at the moments when glacial conditions were the most intense, up to forty times those observed in current conditions.[8] The geochemical composition of desert dusts enables us to identify their source: South America and more specifically Patagonia, where variations in climate have led to a particularly intense aridity and erosion in cold periods.[9] Françoise Vimeux has focused on the analysis of two isotopes present in the ice, deuterium and oxygen 18, whose joint study gives access to the way in which the temperature of ocean regions where the snow of Vostok comes from has varied throughout time. In 1983–84 Michael Bender, then an associate professor of oceanography at the University of Rhode Island, worked at the Centre des Faibles Radioactivités in Gif-sur-Yvette, near Paris. He had thought for some time that oxygen 18 in air bubbles, not ice, should contain a message about the changes in volume of the continental ice and about the sea level during a glacial-interglacial transition. He discussed his idea with Laurent Labeyrie, a paleoceanographer at Gif, and then developed an approach at the Gif laboratory to measure the isotopes of oxygen in air bubbles. The idea attracted Dominique Raynaud in Grenoble, who proposed to prepare samples corresponding to the last climatic transition from the ice core from Dôme C. The results, published in 1985 in *Nature*, demonstrated a link between the oxygen 18 of the bubbles, the deglaciation, and the productivity of the continental and oceanic biosphere. Bender quickly attracted a following for that idea among his graduate students and in his own laboratory with Todd Sowers and at the LSCE in Saclay with Bruno Malaizé, Amaëlle Landais, Gabrielle Dreyfus, and Emilie Capron. It is interesting to note, moreover, that the air trapped in the Antarctic ice, through variations in oxygen 18 of the air and methane, contains information on the evolution of the monsoons that developed in the lower latitudes of the Northern Hemisphere, which in turn has been used to improve the dating of the ice cores of Vostok; this example illustrates perfectly the wealth of ice core archives.

A Huge Lake under the Ice

The 420,000-year-old ice is located at a depth of 3,310 meters, beyond which the core drilling at Vostok continued for more than 300 meters. Signals were then clearly perturbed. The reason for this is the mixing of layers of ice whose correct piling had just been reconstituted to an age of 440,000.[10] Beyond 3,350 meters, a depth reached in 1996, it appeared that all hope was lost to reconstruct a climatic record. The deepest ice then lost much of its interest for our community of paleoclimatologists. Of course, it was still analyzed with as much care. One never knows what one might find. Although in 1999 core drilling was voluntarily halted 120 meters above its surface, the bottom of the drilling held a very nice surprise for us, which turned out to be associated with the existence of Lake Vostok.

We will return later to the discovery and the characteristics of this lake, but imagine our surprise and that of our Russian colleagues when we analyzed the deepest part of the ice core itself. There was nothing very interesting up to 3,538 meters. Then, abruptly, over a few dozen centimeters, the properties changed dramatically.[11] Below that transition, the ice did not contain any gas, very large crystals appeared, from 10 centimeters to even one meter in length, and the electrical conductivity dropped below the threshold of detection of the apparatus. All these properties are characteristic of frozen water from a reservoir of liquid water. This ice thus did not come from precipitations accumulated in the region of Vostok which, after a few hundred thousand years, would have settled into the depths of the Antarctic ice sheet—or at least not directly. The content of isotopes of that very deep ice provided proof of the existence of "lake ice," that is, resulting from the freezing of the lake water. Over less than a meter, we noted a true jump in the concentrations of deuterium and oxygen 18 with very different values above and below 3,538 meters. Not only did the values change but the relationship between the two isotopes did as well; the parameter that we call "deuterium excess" was modified just as abruptly. A mystery at the very beginning, which was easily explained if one admitted that this ice had gradually accumulated, this time from below, following refreezing on the surface of a lake. Located directly below the drilling site, the surface of the lake was expected approximately at a depth of 3,750 meters, and there should have been thus more than 200 meters of lake ice in that location above Lake Vostok.

This was a unique situation, one that could not have been conceived of before it was discovered. The ice sheet, along with the lake, forms an inclined interface creating a difference in level of 600 meters between the Vostok site and the northern part of the lake, where the ice is thicker and the conditions propitious to melting because the temperature at which it takes place drops when the pressure increases—by a few tenths of a degree, but it is enough to make a difference. And since the lake has a constant volume, the quantities of ice that melts to the north are equivalent to those that come out of the refreezing in the southern part. In retrospect, the choice of the Vostok Station site, made at a time when we knew nothing about subglacial lakes, proved to be a very good one. No one at present has yet accessed the water of the lake—which may be the case later if Russian drillers succeed in sampling the ice formed in the Vostok hole from the lake water that surged up when the lake surface was hit and then froze, as reported earlier—but the 85 meters of ice already cored between 3,538 and 3,623 meters were indeed formed from the water that has passed through the lake and has thus retained some of its properties. That ice is available for the scientists of the three countries involved in that project: the United States, France, and Russia. It is a true gold mine for extreme microbiologists, as we will discuss in the last part of this book.

CHAPTER 8

Dôme C

800,000 YEARS AND THE REVOLUTION
OF THE RHYTHM OF GLACIATIONS

When, in 1994, the scientific document that would serve as a foundation for the EPICA project was written, the stated primary objective was to drill in a place that would provide ice older than that found at the Vostok site. The choice of a dome was important because, with the same accumulation, the age of the deepest layers of ice would be older there than at a site located on a flowline, as is the case for Vostok. There is greater thinning of the deep layers, and even if the accumulation at the site chosen at Dôme C was 30% higher than that at Vostok, the age calculated, far enough from the bedrock to limit the risks of perturbation of the ice layers, was estimated at 500,000 years. We thus started with the hope of covering five climatic cycles, twice as many as the drilling at Vostok, which at this time had just barely reached 250,000 years, and we still did not know that the fifth attempt, which was in progress, would prove successful. The EPICA project was then put into motion. A dozen years later we realized, with some satisfaction, that our prediction of 500,000 years was conservative. We have been able to access 800,000 years of archives at Dôme C—twice what Vostok would have provided—but above all, Dôme C has been the entrance into a different world from a climatic point of view.

Antarctica was formed thirty-five million years ago at a time when the opening of passages between Antarctica and Australia, on the one side, and South America, on the other, enabled the formation of a circum-Antarctic ocean current that isolated that continent from the rest of the world. That isolation, probably combined with a decrease in the concentration of carbon dioxide in the atmosphere, led to the formation of an ice sheet that has remained relatively stable since then. The ice sheets of the Northern Hemisphere were formed only much later, around 2.5 million years ago, perhaps there, too, in response to geographical modifications and weaker concentrations of carbon dioxide. Unlike Antarctica, these ice sheets were subject to very marked cyclical variations seen in the records of the volume of ice deduced from the oceanic isotopic data, with a rhythm clearly linked to variations in insolation but which has gradually changed. Up until 900,000 years ago, the glacial-interglacial cycles are less pronounced and were dominated by a periodicity of 41,000 years, probably connected to the variation in obliquity. A second transition, 430,000 years ago, initiated a period characterized by a rhythm of 100,000 years and variations of much greater amplitude, giving birth to those four great glaciations that have been very well documented for the continents, in the oceans, and, thanks to the Vostok core drilling, in the ice of Antarctica. Between these two periods, the many marine records available are a bit murkier because the periodicities linked to the variation in insolation are less clearly identifiable. The EPICA drilling came just in time because it offered the possibility of penetrating into that intermediary zone and, we hoped, of clarifying the picture.

Ice Older than That at Vostok

At Dôme C, the more than 430,000-year-old ice was very deep, below 2,790 meters. We thus had to be patient in accessing it—that depth would be reached only in 2002—but the study of the first kilometers of ice was in itself interesting. For example, it was important to determine whether the variations in temperature such as we reconstructed at Vostok were representative of a larger area of Antarctica. These reconstructions were henceforth available at three different sites because our Japanese colleagues, with whom we were collaborating, had completed their analysis of the cores from Dôme Fuji, located on the other side of East Antarctica, which covered three climatic cycles, from up to 330,000 years ago.[1] Taken as a whole, the three isotopic series indeed gave a very homogeneous image of the climate on Antarctica. For example, all three told us that in Antarctica, during the warm periods that culminated 130,000 and 330,000 years ago, temperatures were as much as 5°C higher than during the Holocene, the period in which we have been living for 11,000 years (Figure 8.1). Another confirmation: the Holocene is the most stable period that Antarctica—and probably the entire planet—has

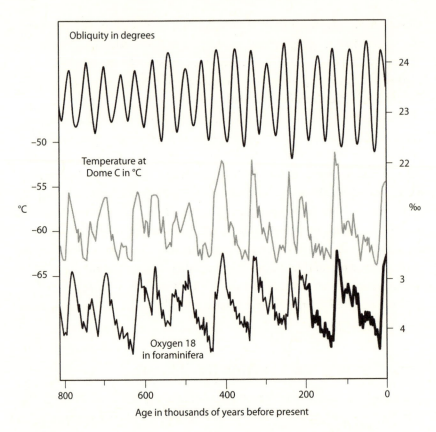

Figure 8.1. Obliquity in degrees. The top curve indicates the level of the Earth's incline; the middle curve, the variation in temperature at Dôme C; and the bottom curve, the concentration of oxygen 18 in the foraminifers, which indicates sea level.

experienced in the last 400,000 years. The two warm periods mentioned above were characterized by a peak of temperature that lasted only around 4,000 years, followed by a cooling period, rapid early on, then slower, before the next glaciation began. And the peak in the intermediary warm period, 240,000 years ago, was also very short. This climatic stability of the Holocene has probably played a role in the evolution and development of our civilizations.

Ice core drilling at Dôme C was continued to the end. The analysis of all the parameters we have discussed—in the ice, in the air trapped within it, in the impurities it contained—was in full swing. A certain number of results were quickly available, including for the deepest ice whose analysis was exciting for whoever wanted to go back in time: imagine, the first 400,000 years were covered by close to three kilometers, and the 400,000 years that preceded them by approximately ten times less. This thinning of the layers near the bedrock nevertheless raised some difficulties. First of all, the existence of perturbations in the last 300 meters of the drilling done at the center of Greenland, GRIP and GISP2, and at Vostok necessarily led us to question the integrity of the equivalent part of the drilling at Dôme C. The risk that these last 10% of the core samples were unusable for climatic ends, as was the case for these three sites, could not be ruled out.

Fortunately that did not occur. One way of demonstrating this was to compare, during a deglaciation, the variations in the concentration of carbon dioxide and methane to that of the temperature. As we had seen at Vostok, these three parameters increased almost at the same time. But synchronous events are not recorded at the same depth in the ice and in the air that was trapped in it. To be convinced of this we have only to look at what was happening in the present. The temperature is determined by the composition of isotopes of the most recent snow right on the surface of the ice sheet, whereas the air, with its current composition, is trapped at the base of the névé at about hundred meters below. As the layers become thinner that "distance" of a hundred meters between the temperature signals and those of the greenhouse gases diminish in depth but remain sufficient to be recorded, unless the layers have been mixed, in which case their properties would be as well. Important variations in all of the parameters are then likely to be measured at the same depth; this was the case in the perturbed zone of the Vostok drilling. Such a situation up to then had not been seen in the drilling at Dôme C, at least as far as 60 meters from the bedrock, a depth at which the deglaciation that occurred a bit less than 800,000 years ago, the ninth as we go back in time, was recorded. We are confident of the integrity of the data provided by the ice of Dôme C over this entire period, which covers eight complete climatic cycles and the preceding deglaciation (Figure 8.1). The situation deteriorates below that level: the variability in all of the indications, whether they are recorded in the ice or in the air, becomes much weaker than what was expected beyond 800,000 years, and everything indicates that a mixture of layers of ice of different origins is the cause of this. We are resigned to abandoning these last 60 meters of ice, at least insofar as information on the evolution of the climate is concerned.

The second difficulty concerns the dating of the ice. As we went deeper into the ice sheet it became increasingly difficult to calculate with precision the way in which the layers thinned and thus to determine the age of the ice using a glaciological model. Fortunately some of the parameters we analyzed were strongly influenced by variations in insolation. The data already available on the Vostok core drilling indicated that the same was true of the content of oxygen 18 in the air influenced by the rhythm of monsoons and thus by precession; Gabrielle Dreyfus in collaboration with our colleagues from Bern extended that record beyond 400,000 years on the oldest part of the EPICA drilling.[2] The drillings at Vostok and Dôme Fuji revealed periodicities in the variations in the oxygen-nitrogen ratio, which most likely resulted from the influence of the local insolation on the metamorphism of the snow in the upper layers of the névé. A similar local insolation signature has also been revealed in the variations in the quantity of air trapped in the ice. We thus have a battery of indicators that enable us, together with the information available on the accumulation of snow and the flow of ice, to propose a viable dating even if it is uncertain by a few thousand years.[3]

Inversion of the Magnetic Field

The Dôme C measurements also offered us, for the first time in an ice core, the possibility of identifying an inversion in the Earth's magnetic field. We all know that the Earth acts as a magnet, that it possesses a magnetic field, and that this is the reason why the needle of a compass lines up following a north-south axis. But it has not always been that way; at certain periods in the Earth's history our compass would have indicated geographic south. This discovery goes back to 1906; it is owed to the French physicist Bernhard Brunhes, then director of the Puy-de-Dôme observatory, who showed that volcanic lava kept the memory of the magnetic field that existed at the time of its formation. By analyzing lava, more or less ancient, he deduced that there were inversions in the Earth's magnetic field, but it would be necessary to wait another fifty years for the chronology of those inversions to be established, showing that the last inversion took place 780,000 years ago. Since

then we have been in the Brunhes period, called normal, with a magnetic field oriented toward the north. Before, the opposite was true, and we must go back more than two million years to rediscover the magnetic field oriented to the north.

Has the ice core sample from Dôme C retained a trace of this inversion that occurred 780,000 years ago? Not directly. In spite of the extreme sensitivity of existing recording methods, there are not enough magnetic particles in the ice for magnetism to be measured. So we must be creative. The Earth's magnetic field forms a sort of shield that protects the Earth from cosmic radiation and affects the rate of production of certain elements—carbon 14, beryllium 10, chloride 36—formed as a result of the interaction between those rays and the molecules present in the upper layers of the atmosphere. When the field disappears, which happens when it is reversed, the production of these elements can be multiplied by a factor of two. Using very detailed analyses, Grant Raisbeck researched the peak of beryllium 10 corresponding to the last inversion that could be located in the ice whose age predicted by the dating model was indeed very close to 780,000 years.[4] We were fortunate because that peak that tells of the magnetic reversal during the so-called Brunhes-Matuyama transition was located only a dozen meters or so above the depth at which the layers of ice are perturbed. But it was a wonderful confirmation: with the ice core drilling at Dôme C we have viable climatic records covering the last 800,000 years, nearly 400,000 more years compared to Vostok.

The verdict, pronounced in 2004 upon seeing the profile of deuterium in the ice analyzed at Saclay, was unambiguous: the rhythm of the climate of Antarctica also changed around 400,000 years ago (Figure 8.1) in a more pronounced way than did the sea level, or continental ice volume, recorded in marine sediments. Earlier the warm periods were less warm, but they lasted longer and, except for the one that happened 650,000 years ago, the glacial periods were less cold. The following year, researchers from Bern and Grenoble provided another important confirmation: the content of carbon dioxide and, to a lesser degree, of methane also varied differently more than 400,000 years ago. The maximal values were also less elevated than those reached since, and the variations in the carbon dioxide varied just as consistently with the temperature as in the 420,000 years at Vostok. The correlation between these two parameters—and this was a very anticipated test, including by skeptics of the greenhouse effect—remained just as strict before that period as after it. The results henceforth covered 800,000 years, and the quality of this relationship between variations in greenhouse effect, dominated by that of carbon dioxide, and climate is truly impressive. One might imagine that the variations in greenhouse effect were at the origin of the change of the climatic rhythm observed at around 400,000 years ago. We can also argue that, just as for a deglaciation, the greenhouse effect participated in that change of climatic rhythm through a modification of the radiative forcing, but the primary cause was likely linked to the modifications in insolation. The insolation received each year at Dôme C presented maxima that, like those of the temperature reconstructed at that site, have been greater since 400,000 years ago than they were before 400,000 years ago (Figure 8.1); proceeding from there to seeing the origin of the change in climatic rhythm observed 400,000 years ago requires only a step.

CHAPTER 9

Rapid Climatic Variations

If we were to rank in importance the climatic phenomena we've been discussing, the existence of rapid climatic variations would be at the top of the list. Who hasn't heard of the halting of the Gulf Stream or, wrongly, of a return to a glacial-like climate that would affect the regions that border the North Atlantic? The media have seized this image and it has even been portrayed in theaters; it is the theme of *The Day after Tomorrow*, a sci-fi thriller from 2005. Let's be clear: this is science-fiction; the icing over of a large part of the United States in a few days is, fortunately, not at all realistic. But the first scenes in which a glaciologist is at work are seductive.

As we have already mentioned, the analyses carried out over the ice in Greenland are at the heart of this notion of rapid climate variation. This idea already had some followers before the first deep ice core drilling took place at Camp Century in northwestern Greenland. But the opinion that was very widely accepted in the first half of the twentieth century was that the climate could only change gradually. That our planet has lived a succession of glacial and interglacial periods was, for many scientists, a concept that was difficult to accept, but to suggest rapid changes was absolutely out of the question, if only because it takes time for a glacial ice sheet to form or to melt.

The First Indications

In the 1930s the study of pollen preserved in peat bogs and lakes in Scandinavia challenged what was believed up to that point: in those regions the end of the last glacial period was clearly marked by oscillations: an initial warming, which ended in a period named Allerod, was followed by a cooling around the period of the Younger Dryas (a name derived from the name of a flower from the Arctic tundra, *Dryas octapetala*). In 1955, a few years after its invention, which earned Willard Frank Libby the Nobel Prize in chemistry, the method of carbon 14 dating was applied to those sediments; it enabled the dating of that oscillation at 12,000 years. Hans Suess, another carbon 14 specialist, confirmed its existence from marine sediments. However, the idea we had at that time of a rapid climate change was quite far from what we understand today since the periods evoked were on the order of a millennium rather than a dozen years. One of the reasons for this is that the carbon 14 dating method was not very precise at that time, indeed on the order of a thousand years or so. And the sediments analyzed, whether continental or oceanic, did not enable us to follow the variations in climate on the scale of decades.

In 1960 Wally Broecker and his colleagues at Lamont near New York were the first to point out the importance of these oscillations, which went from 5° to 10°C. And in the 1970s people began to talk of centennial changes with multiple indications from the point of view of both the archives that provided them—pollen, marine foraminifera, the remains of insects or snail shells—and their geographic distribution. The Younger Dryas was identified in North America as well as in central Europe and in many ocean core samples.

The core drilling at Camp Century provided a great deal of data that support the idea of rapid climate change. In 1969 Willi Dansgaard, Chet Langway, and their colleagues published an ice core record that covered around 100,000 years. At the bottom of the drilling site, they identified ice that could have been formed during the preceding warm period 130,000 years ago. The detailed profile of oxygen 18 content became available in 1972; the recent Allerod/Dryas sequence is very clearly marked in it. More unexpected was the discovery of a series of events, which Dansgaard described as violent, that occurred throughout the glacial period. The comparison with the Byrd records that was done at the time showed a different occurrence in Greenland, visibly very perturbed during the glacial period, and in West Antarctica, which had a much smoother isotopic profile and was thus probably exempt from oscillations of that type.

This increasing number of indications at that time did not draw much attention from the scientific community, which was more interested in the description of conditions that existed in the Last Glacial Maximum and in the establishment of a link between the long-term climate changes and astronomical parameters. During the 1970s there was also speculation about the arrival of the next glacial period, which was thought to be rather imminent—a question of a few centuries, according to some. The fact that the climate could change over such a short period stoked what, in retrospect, appeared to be an unfounded fear that relied on the idea that all interglacial periods are relatively short and on the observation of a cooling that was then taking place in the Arctic. That idea was most likely erroneous, as we have seen, as our interglacial period has every reason to continue over another 20,000 years.

Increasingly Clear Indications

The ice core drilling at Dye3, the first results of which were published in 1982, was an important stage in the history of rapid climate variations. Our Danish colleagues analyzed the results of the drilling in great detail: the current Holocene period covers more than 1,500 meters, and measurements of the oxygen 18 concentration were taken over more than 70,000 ice samples, which enabled the dating of the core year by year over close to 8,000 years. The last climatic transition, represented by close to 50 meters of ice, can be described precisely. Hans Oeschger pointed out that it resembled almost identically a record obtained in the sediments of a Swiss lake, reinforcing the idea that the rapid warming associated with the end of the Younger Dryas resulted from a retreat of waters of polar origin in the North Atlantic with climatic effects somewhat parallel at the center of Greenland and Western Europe. Willi Dansgaard, Jim White, and Sigfus Johnsen[1] examined that transition under a magnifying glass; isotopic analyses showed that this warming, which they determined to be of 7°C, happened in less than fifty years and that the displacement of the polar front was even more rapid, on the order of a dozen years.

Just as important from the point of view of rapid climate variations is the similarity in the climatic curves of Camp Century and Dye3 throughout the last glacial period. Almost all the violent events identified in the former are the same in the latter, in spite of the 1,400 kilometers that separate the two sites. These oscillations were thus not linked to purely local instabilities in the glacial ice sheet; they have at least a regional character. Dansgaard and his team attributed them to changes in the atmospheric circulation in the North Atlantic between two stable states. This explanation was not accepted by

Oeschger and Broecker.[2] Both designated the ocean as the first guilty party, all the more so since Oeschger's team,[3] which participated in the Dye3 project under the leadership of Bernhard Stauffer, discovered the existence of rapid variations in the concentration of carbon dioxide in the air trapped in the ice. Those variations appeared to systematically accompany the abrupt variations in temperature recorded in the ice. In fact, as we will see later in this chapter, we know now that they have been produced by an artifact and hence are not representative of the atmospheric composition in CO_2.

A Connection with Ocean Circulation?

Oeschger, Broecker, and their teams formed the hypothesis that ocean circulation was less active in glacial periods than it is currently. The feeding of the deep ocean occurs from surface waters, but those sink only in well-defined regions of the ocean, where they are sufficiently dense. For that, they must be cold and salty. Today these conditions are encountered only in the North Atlantic, in the Norwegian Sea in particular, and in the south in the Weddell Sea around Antarctica; there is no formation of deep waters in the Pacific Ocean, and the same was probably true in the North Atlantic during a glacial period. Schematically, the ocean would have had two stable states, one corresponding to current conditions, with the production of deep waters in the North Atlantic, the other without. Broecker suggested the possibility of an intermediary mode whose characteristics would be specified a few years later: the surface waters sink more in the south and less deeply.

The events discovered by the Danish scientists corresponded to the passage of one mode of functioning in the ocean to another, which, at the same time, would have explained the variations in the carbon dioxide content, of which the ocean is a huge reservoir, and those of the climate. In fact, the waters that currently sink in the Norwegian Sea are brought there on the surface from tropical regions, and they warm our lands—this is the famous Gulf Stream. When that circulation stops, the arrival of heat is also interrupted, which accentuates the glacial nature of the climate. Broecker quickly became the most ardent defender of the key role of ocean circulation in the oscillations recorded in the ice of Greenland. He also pointed out the role of the atmosphere because the density of the surface waters depends on the amount of freshwater, including that which evaporates and that which is provided by precipitation, the flow of rivers, and the melting of ice. The article he published in 1985 addresses the functioning of this ocean-atmosphere system. A talented scientist, Broecker also knew how to express his ideas to the general public. It is largely to him that we owe the popular image of the conveyor belt with which we associate the Gulf Stream: it circulates on the surface, then the waters that sink into the North Atlantic travel through the deep Atlantic waters from north to south, turn around at Antarctica, bathe the entire Pacific, and then return to the surface.

However, something went wrong. Not so much the idea of the Gulf Stream that stops and starts again—even if the idea was criticized at the time by purists, it has stood the test of time. But the rapid variations in carbon dioxide were troubling, inasmuch as they occurred in the same samples as the climatic variations seen in ice. This should not have been the case: changes in climate and in the composition of the atmosphere that occur at the same time are not recorded at the same depth because the air bubbles are trapped at a hundred meters from the surface. Our Bern colleagues immediately raised the following question: Wouldn't the analyses of carbon dioxide be perturbed by the partial fusion of the ice in the summer or by the greater or lesser amount of carbonates, which when they decomposed in fact produced carbon dioxide? The questions they raised were justified: the ice of Greenland contained enough impurities for the carbon dioxide analyses to be erroneous. Fortunately, those of Antarctica are much purer and are very viable archives of the variations in carbon dioxide in the atmosphere of the past. Forgotten were the variations in the amount of carbon dioxide during the glacial period; there were indeed fluctuations, but they were much smaller and less rapid than those seen at Dye3. However, the isotopic variations recorded in the ice were indeed real, and the core samples of GRIP and GISP2 unquestionably confirmed the rhythm, the rapidity, and the great spikes in temperature that were associated with them.

Confirmation

August 1991: ice core drilling operations were in full force in the center of Greenland at its highest point, the Summit region. At GRIP the Europeans had reached 2,321 meters. They knew they had ice from the last glacial period. A simple calculation showed that it was more than 20,000 years old—

and much older if it was proved that in a cold period the accumulation was smaller than that which currently prevails. Isotopic analyses were moving along. Priority was given to the bottom part of the core. The stated objective: to confirm the existence of rapid climate variations at the end of the last glacial period and the abrupt transition toward current climate conditions, both of which were demonstrated in the ice core of Dye3.

The dating of the GRIP core raises some issues, however. A direct consequence of the relationship with the temperature, the isotopic content of the snow is higher in the summer than in the winter. This seasonal cycle offers a means to date the successive layers year by year. The method could be applied to the ice core of Dye3 over nearly 10,000 years. Further back the seasonal indications faded then disappeared under the effect of diffusion. To our disappointment, this disappearance was faster on the GRIP ice core. Beyond 3,000 years the seasonal isotopic variations were of no help in dating the ice. The great volcanic eruptions, which leave an easily detectable chemical imprint in the ice, were not very useful there because their chronology was increasingly uncertain the further one went back in time. Fortunately chemists took over from there. The content of dust and chemical elements such as calcium, nitrate, and ammonium have a well-marked seasonal cycle. It was work that demanded extreme patience, but the results were there, even if precision decreased with depth. The age of the core was determined within 60 years at 10,000 years and within 800 years at 20,000 years. This ice was located at a bit less than two kilometers in depth, and an age of 40,000 years was thus attributed to the deepest ice, at 2,321 meters.

To reach that goal, the chemists almost continuously perfected methods that enabled them to analyze certain properties of the ice in the field. They worked in a true underground laboratory, which we called the "scientific trench." It was installed under a few meters of snow so that the temperature remained very cold on summer days when the thermometer could climb above zero in the sun. In the trench it was around −15°C, but it was a hive of activity in which about forty people happily worked together. Once the ice core samples, which in general measured two to three meters in length, were extracted, they were carefully inventoried and then stored for a few days to stabilize them from thermal and mechanical disturbances. They were then ready to be examined—one could note with the naked eye layers of ash and even variations in structure that showed annual variations—and to be cut into 55-centimeter pieces that were easily handled and calibrated for the containers that would take them to Copenhagen. The electrical conductivity of the ice was analyzed; it varied depending on the impurities that were present in it. This measurement enabled the scientists to note all of the volcanic eruptions: they are not all visibly seen in a layer of ash. They could also identify precisely the transition between the last glacial period and the current climate. Chemists brought their chromatographers and developed a very slow ice-melting method that enabled them to analyze it on the scale of a centimeter. Others, who had to set up a cave for themselves where the temperature was close to −30°C, studied the physical properties of that ice; for them, observation in polarized light, which revealed the distribution and the size of magnificent crystals, was an essential source of information.

However, some scientists, working for instance on ice isotopes or gases enclosed in ice, were not fortunate enough to have results as soon as the ice was extracted. They had to be patient because the samples had to be brought back to their laboratories. Although it might have been possible to install a mass spectrometer or a line of extraction and gas analysis on the site, it would not have been cost-effective to do so and there would have been the risk of insufficient analytical precision. For the specialists who analyzed the gases and isotopes in the ice, work in the field was limited to cutting and packing the precious samples following specific handling procedures. But we were all impatient to have access to the results. Our Danish colleagues who were isotope specialists did their utmost to make results quickly available; the mass spectrometer in Copenhagen worked at full speed, and a few weeks after the end of the summer campaign we had the detailed profile of the variations in oxygen 18 content for the period from 10,000 to 40,000 years in the past.[4] The isotope specialists told us that we had here a record of variations in temperature at the center of Greenland. We needed to keep in mind that a variation of 1 $/_{00}$ in oxygen 18 corresponded to a change in temperature of 1.5°C — at least that was what we asserted at the time. Two intervals were particularly interesting because they were very perturbed: the deglaciation between 15,000 and 10,000 years ago and the period from 40,000 to 25,000 years ago.

There was a lot of discussion among scientists concerning the age of the events that marked the last deglaciation. Up until 1990, the carbon 14 dating placed the end of the Younger Dryas at 11,000 to 10,000 years ago. The methods that on the GRIP ice core gave access reliable dating, as discussed above, seemed to indicate a greater age. This discrepancy arose from the fact that the rate of carbon 14 production did not remain constant throughout the ages, a difficulty that was avoided thanks to the work of a Franco-American team which, at the same time, was implementing a second method of dating based on the analysis of isotopes of uranium and thorium in their study of a series of coral from Barbados; in 1990 Édouard Bard and his colleagues showed that the carbon 14 ages had to be seriously corrected and that the Younger Dryas ended 11,500 years ago.[5] Confirmation of the end of the Younger Dryas was provided by both European and American glaciologists: the age estimated from the counting of annual layers was 11,550 years at GRIP and 11,640 years at GISP2.[6] The rapidity of the transition was confirmed, whereas the cooling that led to the Younger Dryas was much more gradual. Beyond these aspects, the similarity between the GRIP and the Dye3 records is quite striking.

A few decades appears quite short for a major climate change. And yet the ice in Greenland revealed that certain characteristics of the climate can be modified even more rapidly. Our American colleagues focused on studying in great detail other indications that proved this. At the end of the Younger Dryas it took only five years for the content of dust, a direct proof of the atmospheric circulation, to go from high values of a glacial type to much smaller values characteristic of the current climate. During this transition, which lasted fifty years, the accumulation doubled almost instantaneously in three years and perhaps even from one year to the next.

The similarities observed between GRIP and Dye3 throughout the deglaciation existed during all of the last 40,000 years and erased any doubt that still existed. And at GRIP there was no longer any question of evoking the proximity of the bedrock: 30,000-year-old ice was close to a kilometer away from it. Furthermore, for this ice core we have a sufficiently precise chronology of the sequence of events that occurred in a glacial period. In a few dozen years the climate went from a glacial state to milder intermediary conditions between those of a glacial climate and the current climate. A return to cold conditions was much slower, from 500 to 2,000 years (Figure 9.1).[7] These "sawtooth" sequences were repeated a dozen times. We then estimated that the associated warming was 5–7°C, half of which corresponded to the passage from a glacial climate to the current climate.

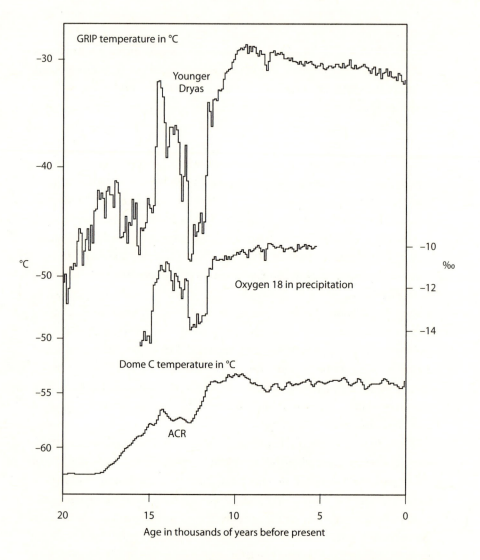

Figure 9.1. GRIP temperature in °C. The isotopic record (oxygen 18 content of the ice) obtained from the GRIP ice core drilling for the last 20,000 years reveals two extremely rapid warmings, one around 14,500 years ago, the other around 11,500 years ago (upper curve). These same variations are recorded in the lake sediments of Lake Ammersee in Bavaria, indicating that these rapid variations also affected western Europe (middle curve), while over that same period variations were much more smoothed in Antarctica (bottom curve). ACR: Antarctic Cold Reversal.

Rapid Events during a Warm Period?

July 12, 1992: GRIP core drilling reached the bedrock at a depth of a bit more than three kilometers. A preliminary dating allowed us to hope that those 3,000 meters of ice would cover 250,000 years. Up to around 100,000 years ago, the isotopic profile revealed an entire series of rapid and large variations that were completely in line with those already identified in the glacial period after 40,000 years ago. This was not surprising because around 100,000 years ago we were still in that glacial period. We counted twenty-four of those events to which, a few years later, Wally Broecker gave the name Dansgaard-Oeschger, in homage to the key role played by those two researchers.

In the field, using different analyses, it was easy to estimate the approximate age of the ice as the drilling progressed. At a depth of 2,800 meters, all indications suggested that we were entering into the Eemian, the warm period that preceded our own. This was around 110,000 years ago. This interglacial ice was close to 100 meters thick. As the core samples were extracted, we felt we were in for some surprises: the dielectric conductivity did not present the stability characteristic of the ice from the last 10,000 years, and the size of the crystals and various proxies showing the contribution of continental aerosols provided evidence of rapid variations. The isotopic curve was thus impatiently awaited. It was available in the month of October; there were indeed many surprises.

The isotopic curve revealed the existence of rapid—indeed catastrophic—events during the last interglacial period, the Eemian, which began around 135,000 years ago and ended around 115,000 years ago (Figure 9.2).[8] No one expected that this period, which was slightly warmer than the current climate, would be completely different. But it was interrupted several times by shifts toward conditions rather close to those of a glacial period. Transitions were very rapid (a few dozen years). Cold conditions persisted, depending on the case, for periods of 70 to 5,000 years. This was the first time that such transitions were observed during that period, which had sometimes been considered analogous to the climate of our planet in terms of climatic warming due to an increase in greenhouse gases. The stability of the current climate, which has prevailed for around 10,000 years, thus appeared to be unusual.[9] This was an enormous surprise for specialists in climates of the past, especially since the detailed analysis of one of these rapid variations yielded

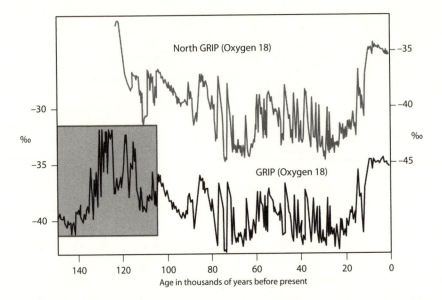

Figure 9.2. Variation in the oxygen 18 content along the GRIP and North GRIP ice cores; beyond 105,000 years the recording of the GRIP cores was perturbed as a consequence of a mixing of layers linked to the proximity of the bedrock.

completely unexpected results. A cooling estimated at 14°C, which took place over the course of twenty years, lasted only seventy years and was followed by a return to a warm climate, also over the course of twenty years. And, simultaneously, there were just as abrupt and spectacular modifications of different indicators of atmospheric circulation that showed drastic changes in the atmospheric circulation and supported the image of a shift toward a glacial climate right in the middle of a warm period.[10]

These results raised many questions, the first of which concerned their viability. Wouldn't there have been a mixing of different layers of ice? Was the isotopic content of the ice indeed representative of the temperature of the site and its chemical composition of that of the atmosphere? To address all of these questions the GRIP team put forward solid arguments and quickly published these results. And even though we were cautious, we did propose an erroneous interpretation in *Nature*. There were real instabilities in the GRIP records, but we had believed then (wrongly) that we could interpret them as climatic variations occurring in a warm period. We were disappointed to learn not too much later that they were in fact the result of a "common" blending of layers of ice linked to the proximity of the bedrock.

The warning came from our American colleagues who had less luck than we did in their drilling operation. Their new drill, of impressive size, allowed them to extract core samples 6 meters long and of greater diameter (12 centimeters versus 10 with Istuk). This drill was entirely satisfactory, but the cable was a source of problems. As already reported in chapter 6, the 1992 season had to be interrupted at a depth of 2,200 meters, and suddenly our colleagues were lagging. A new cable was brought to the GISP2 site in May 1993. The Americans reached the base at the beginning of July, one year after the Europeans, whose record—with an ice core at 3,054 meters—they beat. In the meantime we published the GRIP results, but it was with some anxiety that we awaited confirmation of this surprising structure of the last interglacial period.

The results of the GISP2 drilling[11] indeed raised some serious questions about the validity of our interpretation. Beyond 100,000 years ago the records obtained from each of the drillings began to diverge, indicating that the stratigraphy of the cores, or at least of one of them, was modified due to distortions linked to the flow of the ice close to the bedrock. The presence of inclined layers, generally associated with the phenomenon of folding, was observed at GRIP only at 130,000 years ago, whereas it was noticed much earlier at GISP2, leaving a slim hope that only that second drilling was perturbed. But analysis of the composition of air bubbles will definitively reject the idea of rapid climatic variations during the Eemian. Let's take, for example, methane, which mixes rapidly in the atmosphere; in the absence of perturbations, we should, for a given period, obtain a record that is practically identical in Greenland and in Antarctica. But beyond 100,000 years, this was not the case at either GRIP or GISP2: neither record could be compared to that then available at Vostok, which we knew had not been perturbed because the corresponding ice sequence was more than 1,500 meters above the base. We had thus gone too far in the interpretation of the rapid variations recorded at GRIP during the Eemian. They were unquestionably the result of a blending of ice layers formed at different periods in the past; in no way did they testify to rapid climate changes.

A striking proof of this was provided by the results obtained at North GRIP (Figure 9.2), which indicate that the deepest layers remained in chronological order, probably because the flow of the ice was facilitated there by the presence of liquid water interfacing with the bedrock. This ice core sample, whose age was estimated at 123,000 years, covers part of the Eemian but shows over that period no change of the type recorded at GRIP and GISP2. At most it reveals a new Dansgaard-Oeschger event, the twenty-fifth, around 110,000 years ago, at the moment when the last glaciation was beginning. Everything was in order: the preceding interglacial period was indeed a climatically stable period, at least compared to the last glacial period, which throughout (or almost throughout) its 80,000 years was very agitated. Indeed, between 100,000 years ago and the Last Glacial Maximum 20,000 years ago, isotopic variations recorded at GISP2 are exactly the same as those revealed a year earlier by the ice of GRIP. This comparison held just as true for the North GRIP drilling, and these three isotopic records are also very similar over the entire last deglaciation, which leads us to the beginning of the current Holocene period, which began around 12,000 years ago.

Initially Underestimated Changes in Temperature

However, it quickly became apparent that the way in which we were evaluating the variations in temperature from the isotopic composition of the ice in Antarctica led us to underestimate it in Greenland. The warning signal was sounded by our Danish and American colleagues, who at GRIP and GISP2 had measured the temperature along the bore holes extremely precisely and meticulously. The diffusion of heat was sufficiently slow for this temperature profile to partially keep the memory of conditions that ruled on the surface in the past. Mathematical methods then enabled us to estimate the mean temperature for periods such as the Last Glacial Maximum. And then a new surprise: in the last 20,000 years the climate at the center of Greenland has warmed by close to 25°C, or more than twice the 12°C that we had announced using the isotopic composition of the ice.[12]

What happened to the rapid warming associated with the Dansgaard-Oeschger events? Was their amplitude also underestimated by a factor of two? The temperature profiles measured in the bore holes could not answer that question because they involved events that were too short to be kept in the archives. Here, too, we had to shift gears. But thanks to an American team—that of Jeff Severinghaus in San Diego—that rapidity became an ally.[13] We have already noted this: air bubbles are definitively trapped in the ice at the base of the firn at about a hundred meters in depth. The snow is packed under its own weight, becoming less and less porous the deeper it gets until it becomes completely airtight. Imagine that a rapid warming occurred on the surface. The effect would gradually travel into the firn but take dozens of years for it to be completely felt at a hundred meters in depth. During that time, it would be warmer on the surface than at the base of the firn, whose depth would change slightly. And in such a porous medium there would be a "fractionation" of the gaseous compounds due to gravity on the one hand and to that difference in temperature on the other. Severinghaus was interested in the nitrogen isotopes and in those of argon whose proportions had no other cause to be modified except by the processes that took place in the firn. As theoretically expected, those isotopic compositions changed measurably each time the temperature varied rapidly on the surface. A new thermometer was born. It enabled the evaluation of the amplitude of the successive warmings, which are systematically greater than those estimated from the isotopic analysis of the ice and could reach as much as 16°C. Such a high value was surprising, but it appeared very viable thanks to analysis done independently at Bern[14] and by Amaelle Landais at Saclay.[15]

The Connection with the Ocean Henceforth Demonstrated

Fifteen years went by, during which an interest in these rapid variations—their geographic extent, the mechanisms that were at their origin, the possibility that they would occur in the coming centuries—continued to grow. In fact, in the 1980s specialists in marine sediments weighed in. If the seductive scenario of a connection between the violent changes recorded at Dye3 and changes in the ocean currents in the North Atlantic proposed by Oeschger and Broecker a dozen years earlier were true, we should have found a trace of those events in marine sediments. The paleoceanographers got to work. The first indications that this was indeed the case were revealed by sedimentological analysis. Some core samples were interrupted by a series of layers that came from the base of the Canadian continent. They were known as "Heinrich layers," after the German scientist who discovered them. And Gerry Bond, a colleague of Broecker at Lamont, demonstrated the cyclical variations in the color of the sediment. These teams and other researchers joined together and in 1993 proposed a coherent image of these different observations.[16]

Six Heinrich layers were identified between 70,000 and 14,000 years ago and were mapped. Their continental origin was confirmed. Furthermore, a decrease in the amount of oxygen 18 in the seawater was associated with them. These two elements suggested that they were connected to massive discharges of icebergs (poor in oxygen 18) breaking off from the North American ice sheet when it grew to the point of instability. These massive detachments occurred at the end of the cooling phase, a period that lasted 5,000 to 10,000 years. We thus find again the "sawtooth" structure of the ice of Greenland, and it was tempting to associate the events identified in the marine sediments of the North Atlantic with the most marked rapid changes recorded in the ice. In fact, this correspondence has been demonstrated by examining in detail the different marine and ice core records. There was, through the ocean, a connection between the massive discharges of icebergs and at least some of the rapid variations recorded in Greenland. It has been proposed that these discharges, whose volume could be on the order of a million cubic kilometers—or the equivalent of nearly two kilometers of water over all of France—were produced at a time when it was very cold in the North Atlantic and when the Gulf Stream was practically stopped. The discharges were conveyed by a retreat of the edges of the glacial ice sheet toward the interior of the continent, an ice sheet that for a few hundred years provided much less freshwater to the ocean. The surface waters became a bit denser, and that would have been sufficient for the sinking of the waters of the Gulf Stream to start up again in the North Atlantic with, as a consequence, a rapid warming of those regions, Greenland in particular.

The relationships between the climate of Greenland and the conditions that existed in the North Atlantic are even closer than these initial comparisons might allow us to see. Paleoceanographers have since then undertaken increasingly detailed analyses over different parameters (temperature, salinity, ocean circulation), which have revealed that each of the Dansgaard-Oeschger events was systematically associated with changes in these properties. But the atmosphere was also involved. This is seen in the extremely rapid variations of the quantity of dust present in the Greenland ice; it was very elevated during the cold phases that preceded warming but 10 to 100 times

less once that warming began. These differences correlate with drastic modifications in the winds. There were, moreover, true upsets that accompanied each rapid event both in the atmosphere and on the surface of the North Atlantic, inasmuch as the oceanic source of the water vapor, from which the snow that fell at the center of Greenland was formed, also moved rapidly. A study conducted by Valérie Masson-Delmotte from the combined analysis of deuterium and oxygen 18 at North GRIP has clearly demonstrated the conditions prevailing at the surface of the ocean.[17] At first sight, the concentrations of these two isotopes in the snow vary in a parallel fashion. But if we look closer, slight differences come to light, which can be used to identify the conditions that prevailed on the ocean surface in regions from which masses of air, which arrived on Greenland, originated. Jorgen Peder Steffensen and his colleagues have shown that these conditions can change very abruptly during periods of rapid warmings, with shifts of the atmospheric circulation in the Northern Hemisphere resulting in changes of 2–4°C in the temperature of the Greenland moisture source from one year to the next.[18]

In France we are very aware of the influence of the weather that exists in the North Atlantic on the climate of a good part of our country and more generally over Western Europe. This is simple good sense even without the available data, which would suggest that the upsets characteristic of the glacial climate of the Atlantic were noticed by our ancestors. Records exist and confirm that intuition. Furthermore, the Dansgaard-Oeschger events were accompanied by clear modifications in precipitation, which was less abundant in periods of intense cold than during milder phases. The repercussions on vegetation are obvious and were reconstructed through an analysis of pollen in continental series like those of the Grande Pile in the Vosges and the Échets near Lyon, as well as in marine core samples taken near the coasts. The isotopic composition of the carbon of the stalagmites in our caves in the Massif Central, formed of calcite ($Ca\ CO_3$), was influenced by that of the vegetation that covered the ground and retains a record of it, while that of its oxygen reflects the isotopic composition of precipitation. Stalagmites (we have an example of this in the Villars cave in Dordogne) are thus excellent records of Dansgaard-Oeschger events that, thanks to isotopes of uranium and thorium, are generally able to be dated more precisely than is possible from polar ice or marine sediment.

Consequences on a Planetary Scale

These rapid variations were also identifiable on the other side of the ocean in the United States and as far as Brazil. But it took only a few months after the extraction of the core sample at GRIP for us to realize that the rapid climatic variations had repercussions beyond these regions close to the Atlantic: Western Europe, America, and Africa. We owe this to the work carried out by Jérôme Chappellaz; as soon as the first part of the GRIP core sample, covering the last 45,000 years, was available, it showed that a significant increase in the content of methane in the air bubbles extracted from the ice generally corresponded to these variations. The production of methane is connected to the extent of flooded zones. Consequently, these increases likely prove variations in the continental hydrological cycle at low latitudes, which suggests that the rapid events influenced the climate of the Northern Hemisphere overall. The variations in methane have henceforth been documented over all of the glacial period. What is more, the analysis of stalagmites taken from Chinese caves enabled scientists to establish a precise connection between the Dansgaard-Oeschger events and the rhythm of monsoons, which were more active at the time when the temperature in Greenland was relatively mild and the concentrations of methane were high. The circle was closed. It became very clear that the entire Northern Hemisphere was perturbed during the last glacial period: temperatures, precipitation, winds, the intensity of the monsoons—all those parameters seem to have varied more or less simultaneously, seemingly over the entire hemisphere. Paleoceanographers also discovered traces of these rapid events in the Baltic Sea, in the Mediterranean, in the Indian Ocean, and even in the distant Pacific.

But what about the Southern Hemisphere? In Brazil, as we've just mentioned, changes associated with the Dansgaard-Oeschger events were detectable but not very marked. They were also relatively difficult to identify in the marine sediments of the Southern Hemisphere. In fact, it was the ice core samples from Antarctica that provided the most detailed information. Granted, the absolute age has not been precisely established there, but the ice of Antarctica and Greenland can be placed on a common scale of time thanks to successive peaks in methane recorded in the north and in the south. Thanks to the Byrd core samples and, more recently, to those of Dronning

Maud Land and Dôme C, we now have a good idea of what occurred in Antarctica during the glacial period. Each Dansgaard-Oeschger event had a counterpart in Antarctica but of a different form—the phase of warming and that of cooling were relatively symmetrical there, whereas in the north the warming was rapid and the cooling slower—and of lesser intensity.[19] The events there did not exceed 2–3°C, while it reached as much as 16°C in Greenland. The south began to warm up when the north was very cold, and the rapid warming occurred in Greenland at about the time when the temperature reached its maximum in Antarctica. Thus, for the most intense events, warming could have begun there, in Antarctica, more than 1,000 years before Greenland began to warm.

This structure, which is sometimes erroneously called a "seesaw" and can also be used for the last deglaciation, clearly proves the key role of the Atlantic Ocean and its circulation in the transfer of climatic variations between each of the hemispheres, but many questions remain. Is the south a simple "slave" to what occurred in the north, the absence of thermohaline circulation in a cold period favoring the gradual warming of the south and its resumption during the rapid warming of Greenland provoking a cooling in the south?[20] On the contrary, was the south, as might be suggested by the fact that the warming started earlier there, a true agent through changes in winds from the west or from the ocean circulation in the Southern Ocean or even from the injection of freshwater coming from Antarctica? Finally what about the role of tropical regions? Some work suggests that a change in rhythm of the El Niño phenomenon would be likely to modify the amount of freshwater in the Atlantic and as a consequence the thermohaline circulation. The abrupt and intense nature of the variations recorded in Greenland may suggest that mechanisms that took place in the North Atlantic were at the origin of the very great variability of the climate observed during the glacial period, at least for the most part. Despite the progress made in recent years, a progress to which the data acquired on the core sample of Greenland and the Antarctica have greatly contributed, no one has a definitive answer for now. For the debate to be settled, it would be necessary for inherently complex models to account for the interactions among the atmosphere, the ocean, and the polar ice caps on these scales of time. A lot of ground remains to be covered.

CHAPTER 10

The Last 10,000 Years

AN ALMOST STABLE CLIMATE

The calm after the storm: this is the image we have from the climatic records of polar regions. In Antarctica the contrast between the hills and the valleys that followed each other during the last glacial period, then the deglaciation, and the stability of those same records since the beginning of the Holocene, a bit more than 10,000 years ago, is clear. And above all, there is nothing in common between the highs and the lows, rapid warming and slower cooling, which punctuated the temperature variations in Greenland between 100,000 years ago and the end of the Younger Dryas, 11,500 years ago, and the "flatness" of the same records since then. Climatically speaking, they are two different worlds.

Not entirely, however. Around 10,000 years ago we were already in an interglacial period—temperatures were even a bit warmer than they are today, at least in Antarctica, but our planet was not yet completely free of the jolts characteristic of colder periods. The reason for this is the slow pace at which the enormous glacial ice sheet that covered the northern part of North American disappeared. A good third of it still existed 10,000 years ago. Mild temperatures favored the melting of the ice, which sometimes led to the formation of "glacial" lakes imprisoned by ice barriers. This was the case 8,200 years ago in northeastern Canada.[1] But the barrier that closed Lakes Agassiz and Ojibway gave way, freeing enormous quantities of freshwater, on the order of 200,000 km³, which poured into the North Atlantic in less than a hundred years. The consequences of a stop in the thermohaline circulation that followed were very widely felt in the Northern Hemisphere, in the Atlantic Ocean, in Western Europe, in Africa, and as far as the regions affected by monsoons. This is also seen in the decrease in the concentration of methane analyzed in the ice of Greenland in which the chronicle of that event is most faithfully recorded. The climate cooled there by around 5°C in a few dozen years; this colder climate, characterized by a decrease in precipitation, continued for about seventy years, and the return to earlier conditions again took a few decades.

Other events, considered abrupt, have marked the Holocene period, with rapid variations in the covering of sea ice and in precipitation in the high latitudes of the hemisphere 4,000 to 5,000 years ago. Then there was rapid cooling in western Europe, periodic droughts over a large part of North America, and notable climate changes in South America. However, the most remarkable event remains that which occurred 8,200 years ago because in the second part of the Holocene, there were no longer ice sheets on North America likely to quickly free large quantities of freshwater.

Throughout the Holocene, conditions of insolation changed in response to variations in the Earth's orbit. Nine thousand years ago summer insolation, as well as the contrast between summer and winter, reached a maximum in the Northern Hemisphere and remained high until the middle of the Holocene. This strong seasonal contrast increased the difference in temperature between the continent, which warmed or cooled rapidly, and the ocean, which had difficulty following because of its great thermal inertia; as a consequence, monsoons in the tropical regions intensified. This was observed from the beginning to the middle of the Holocene in North Africa, India, and Southeast Asia. The seasonal variations of insolation would also be at the origin of the gradual intensification of the rhythm and the intensity of the El Niño phenomenon, which was less marked at the beginning of the Holocene than in the second part. Finally, and we now return to the ice, strong summer insolations could be at the origin of the weak expansion—indeed the absence—of many glaciers from the Northern Hemisphere, which did not begin to expand until around 5,000 years ago.

Volcanism and Solar Activity: Natural Climatic Forcings

As we get closer to the current period, our knowledge of the evolution of the climate becomes increasingly detailed, both in time and in space. But just as much as the most precise description possible of the evolution of the climate, it is important to look carefully at the phenomena that could be at the origin of it, to specify what we call climatic forcing. In this period of relative stability, certain forcing, which retreats from the forefront when we are looking at the great changes of the last glacial period, must be taken into consideration. This is the case for volcanic eruptions and variations in solar activity, forcings that are recorded in polar ice.

The chemical composition of the ice was modified at each sufficiently large volcanic eruption with, for example, an increase in the content of sulfates. The eruptions in the Northern Hemisphere are clearly recorded in the ice of Greenland, those of the Southern Hemisphere in the Antarctic ice. Some eruptions occurring close to the equator or in the tropics can leave traces in both the north and the south when they are very violent. Beyond the historical archives, polar ice sheets thus enable us to establish a calendar of eruptions that have marked our recent and more distant past, a calendar that in some cases proves more precise than that kept in historical archives. For a long time it was believed that the eruption of the volcano of Santorini, in Greece, whose climatic consequences were felt as far as China and California, occurred in the sixteenth century; the ice in Greenland, dated year by year, place that eruption a century earlier; it is that chronology, confirmed by other methods, that is now considered fact.

The influence of these volcanic eruptions on the climate results from the presence of ash, which darkens the atmosphere for a few weeks, but above all from the formation of sulfur aerosols that result from the oxidation of gaseous compounds emitted during the eruption. These aerosols have an effect over longer periods, as much as several years. Volcanoes are thus related to cooling and, for a limited time, they can, as a result of an increase in the optical thickness of the atmosphere, have negative forcings equivalent to that associated with an increase in the greenhouse effect of anthropogenic origin. This forcing was on the order of 3 Wm^{-2} for the eruptions of Krakatoa in 1883 and of Pinatubo in 1991 and weaker (2 Wm^{-2}) for those of Agung in 1963–64 and El Chichon in 1982, all of which are relatively well documented. It is more difficult to estimate the variations in optical thickness for eruptions only recorded in the ice as well as the geographic distribution and the duration of associated perturbations, but this is the only approach available for going back in time. This effort of reconstruction has concentrated on the last millennium; with more than 10 Wm^{-2}, the prize goes to the eruption of 1259, to which no named volcano has up to now been associated. Next in line are the eruptions of Laki in 1783–84 and Tambora in 1815 (both around 5 Wm^{-2}), which, a year later, led to the year without summer. We can imagine that the year 1260 did not have a summer either.

The climatic role of variations in solar activity is the subject of much debate. Some scientists suggest that the Sun is at the origin of the rather regular rhythm of the rapid variations of the last glacial period, which are separated by a duration close to 1,500 years or by a multiple of that duration; others see the imprint of the Sun in some records covering the Holocene in which that same periodicity, generally fleetingly, is present. As we will see, we cannot attribute the recent warming to a variation in solar activity, but it is undeniable that those variations have an influence on our climate; thus they could be at the origin of the Little Ice Age that affected Europe between the fifteenth and nineteenth centuries. Whatever the case, we understand the importance of knowing the past variations in solar activity with sufficient precision. Here, too, polar ice is of precious help because solar activity has been precisely measured for only a few decades, and the observations, for example, those of sunspots, which enable us to estimate such activity, were not available before 1610. Cosmogenic isotopes took over, formed, like carbon 14 or beryllium 10, through the action of the cosmic rays on the upper layers of the atmosphere. Their rate of production depends, among other parameters, on solar activity. Beryllium 10, which can be analyzed in polar ice, is particularly useful for reconstructing this activity over millennia because concentrations of carbon 14 are also influenced by different processes intervening throughout the carbon cycle.[2] There is a lot of agreement among the estimates based on the observation of sunspots and those deduced from beryllium 10. There is, however, a serious problem of calibration that can cause those estimates to vary by a factor of 4 at least. The most recent work indicates that the variations in solar activity are weaker than anticipated a few years ago; its increase between the Maunder Minimum, in the seventeenth century, and the current period would not exceed 0.1%, or in terms of climatic forcing, less than 0.2 Wm^{-2}.

How Long Has Human Activity Been Changing the Composition of the Atmosphere?

Another debate is the one that surrounds the variations in concentrations of carbon dioxide and methane throughout the Holocene. Everyone agrees regarding the data that are viable and very well documented: the concentration of carbon dioxide slightly diminished until 8,000 years ago, by 7 parts per million by volume (ppmv), then increased by 20 ppmv between that period and the beginning of the industrial revolution, while that of methane diminished until 6,000 years ago, from 730 to 580 parts per billion by volume (ppbv), then increased again to a level of 730 ppbv, also at the beginning of the industrial revolution. The variations in nitrous oxide are parallel to those observed for carbon dioxide. In terms of radiative forcing, the variability of greenhouse gas effect over this entire period was 0.5 Wm^{-2}. This represents around 20% of the increase observed in the last 200 years in response to the emissions resulting from human activity. The American researcher Bill Ruddiman has posed the hypothesis that the increases observed—in the last 8,000 years for carbon dioxide and 6,000 years for methane—are due to those activities. This hypothesis is rightly controversial, as human activity, in particular at the level of deforestation, does not appear sufficiently great to lead to the increases observed.

Once again, it was arguments based on the analysis of air bubbles trapped in the polar ice that were decisive: the variations in carbon 13 from carbon dioxide, as revealed in the Antarctic ice for the Holocene, are not compatible with the idea advanced by Ruddiman.[3] Ruddiman built his argument around the data from the Vostok ice cores, the only ones available when he defended his hypothesis in December 2003 during the fall meeting of the American Geophysical Union, which was held in San Francisco. He compared what happened in the last 10,000 years with the data available for the three warmest periods of the preceding interglacial periods, which were as warm or warmer than the Holocene and that mark the beginning of each new climatic cycle. The last interglacial period and the two that preceded it began around 130,000, 245,000, and 335,000 years ago, with warm periods for which, given the similarities between the climatic records deduced from ice cores extracted in Antarctica on the one hand and the records obtained from marine sediments on the other, the community of glaciologists adopted the terminology defined by paleoceanographers, Marine Isotopic Stage (MIS) 5.5, 7.3, and 9.3, respectively. Ruddiman noted that the highest concentrations of carbon dioxide and methane were observed at the beginning of those three warm periods—MIS 5.5, 7.3, 9.3, and 11.3—then decreased. Only the Holocene has escaped that trend; this is the argument Ruddiman used to see the hand of man in the increase in concentrations of those two greenhouse gases in the last 8,000 years for the former and 6,000 years for the latter.

The full length of the preceding interglacial period, MIS 11.3, which the marine records indicate began around 420,000 years ago, was not recorded, at least continuously, in the Vostok ice. The EPICA ice core drilled at Dôme C covered all of the 11.3 stage, and the data obtained from those samples would disprove Ruddiman's hypothesis two years later. Moreover, in the meantime, it was possible to extend the records from Vostok by 20,000 additional years by reconstructing a correct chronology of the corresponding layers of ice affected by a process of reversal due to the ice flow disturbance occurring at very great depth. All of stage 11.3 and of the transition that preceded it were thus indeed present at Vostok as was confirmed by the comparison with all of the data available from the EPICA drilling. What was observed there? Let's first take the case of methane. The highest levels were observed at the very beginning of the interglacial period, then, rather similarly for the Holocene, decreased for 5,000 years, and then began to increase. This increase could obviously not be attributed to human activity, which was practically nonexistent 400,000 years ago, and it weakened Ruddiman's argument, which was also not obviously supported by the records of concentrations of carbon dioxide recorded at Vostok and at Dôme C throughout that interglacial period.

Why was this counterexample provided to us by stage 11.3 so crucial? After all, over the four preceding interglacial periods, after the initial decrease, only stage 11.3 showed an increase in methane, but it was then necessarily of natural origin. In fact, among these four preceding interglacial periods the most important one for comparison with the Holocene is indeed stage 11.3 because it is the one for which variations in astronomical parameters are most directly comparable. The reason for it is that we have been, for tens of thousands of years, in a period of weak eccentricity during which the Earth's orbit is round, or almost. The direct consequence of this is that the variations in insolation received in the summer in the Northern Hemisphere, that which according to the astronomical theory governs the rhythm of glaciations, are little marked: we must indeed go back 400,000 years to find similar astronomical conditions.

These exceptional conditions could be a true piece of luck in forecasting our future. None of the warm periods of the interglacial stages 5.5, 7.3, and 9.3 lasted more than 10,000 years. And since our Holocene has already lasted for 10,000 years, we should ask questions about the rapid arrival of the next glaciation, which would mark the end of our interglacial period. Ruddiman pushes his reasoning in that direction: not only does he propose—and we have seen the weakness of this—the hypothesis that the increase in the greenhouse effect observed since the first half of the Holocene is mainly linked to human activity, but he argues that without that increase we would be, let's say naturally, on the verge of entering into a glacial period or even already in the process of undergoing the first consequences of one.

The full story is not yet clear, but the reason the Holocene would be continuing in the absence of the human impact could well be related to the astronomical configuration of weak eccentricity from which we are currently benefiting and which would continue for several more tens of thousands of years. It isn't possible over such long periods to use models that represent the behavior of the atmosphere, the ocean, and the polar ice caps in a detailed way. Such an exercise is in any case beyond our reach regarding the available means of calculation. Modelers can nevertheless project into that very long-term future by relying on either very simple conceptual models or intermediate complex models. None of these projections foretells an entrance into the next glacial period before several tens of thousands of years.

These projections, based on the perfectly known evolution of astronomical parameters, are remarkably corroborated by the available data on the closest stage 11.3, as we have said, of the current period and of our future in terms of variations of insolation. Marine sediments attest that this warm period has been exceptionally long compared to the three last interglacial periods, and the data from the ice cores at Dôme C provide a remarkable confirmation of this. The amount of time that the temperature has been the same or above that which on average the Holocene has known is close to 30,000 years.

Everything thus indicates that the next glacial period will not be anytime soon. This is not the problem that future generations must face; rather it is that of the warming put into motion by an increase in the greenhouse effect linked to human activity since the beginning of the industrial era around two hundred years ago and its increase during the twenty-first century and beyond.

PART THREE

THE WHITE PLANET TOMORROW

CHAPTER 11

The Climate and Greenhouse Gases

Through the eyes of glaciologists we have led you to the discovery of our white planet, of the world of ice with such variable shapes and with such a rich memory. What will become of this ice on a planet whose climatic history is no longer written by Mother Nature but quite probably is already influenced by the activities of humans and which will probably be even more so in the decades and centuries to come? This question, which bears on the role of human activities in the warming we have been experiencing for a few decades and on the evolution of our climate from now until the end of the century and beyond, is the focus of the next several chapters. But let's first go back to the greenhouse effect, to its role in the climatic machine, and to its evolution during the last centuries in response to human activity.

The Greenhouse Effect: A Truly Beneficial
Natural Phenomenon

To explain the greenhouse effect we must look at it as climatologists, which will enable us to say a bit more about the functioning of our climate, of which the Sun is the initial driving force. For even if the Earth absorbs only a small part of the energy the Sun emits, that energy is much greater (around 7,000 times) than the geothermal flux that comes from inside the Earth. That flux thus has no notable influence on the average climate of our planet.

In a given place, the solar energy received at the top of the atmosphere—insolation—is extremely variable. It follows the rhythm of the succession of days and nights, which is connected to the rotation of the Earth on its axis, and of that of the seasons, which result from the obliquity, the tilt of that axis in relation to the plane of the Earth's orbit. It also depends on the average incline of the Sun's rays and thus on latitude. The result of these diurnal, seasonal, and geographic variations: over an entire year, the average insolation received at the top of the atmosphere is indeed much less than the energy that would go through that same surface if it were placed permanently perpendicular to the Earth-Sun direction. This latter, which we designate wrongly by the name of "solar constant," is close to 1,365 Wm^{-2}, whereas the average energy received, approximately 342 Wm^{-2}, is only a fourth of that. This ratio owes nothing to chance but to the fact that the surface of a sphere of radius R ($4\pi R^2$) is four times that of a circle passing in its center (πR^2). This is, quite simply, the way in which our Earth is viewed from the Sun.

On a longer scale of time other factors come into play. There are astronomical parameters (eccentricity, obliquity, and precession), which we have already mentioned. They enable us to calculate the insolation at a given time in the past, present, or future, as a function of latitude. These astronomic parameters vary, as we have seen, very slowly, but another factor must be taken into account. This is the intensity of the heating. In fact, the solar constant varies slightly with the rhythm of the fluctuations of the activity of the Sun, which is seen, for example, in the number of sunspots.

The variations in solar activity have probably been great since the beginning of the history of the Earth, a period when the power of the Sun was around 30% weaker than it is today. Those variations do not, however, have any influence on the time scales of the great climatic cycles characteristic of the Quaternary and are thus legitimately ignored. But solar activity also varies over short periods of time with a very marked cycle of eleven years over which are superimposed periodic variations such as the one characteristic of the period of the Maunder Minimum, which corresponds to a near absence of sunspots during the second part of the seventeenth century. These variations in solar activity are weak, on the order of 1‰ (or 1 per mil) for the eleven-year cycles and a bit more than 3‰ (or 3 per mil) at the most, since that Maunder Minimum. Their influence on our climate nevertheless is the object of an intense debate that we will discuss in the next chapter.

Up to now, we have looked at the energy that arrives at the top of the atmosphere. But as we see in figure 11.1, which is very simplified, many things occur within the atmosphere itself, most of the mass of which is concentrated over approximately twenty kilometers of thickness. Around 107 Wm^{-2}—or approximately 30% of the 342 Wm^{-2} that, on average, arrive at the top of the atmosphere—are reflected back to space. Two-thirds of this is due to the

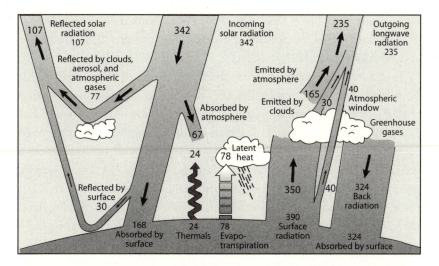

Figure 11.1. How the greenhouse effect works. Source: IPCC, *Climate Change 2007: Fourth Assessment Report* (Cambridge University Press, 2007). Note: All numbers correspond to Wm⁻².

presence of clouds and the rest to the diffusion through air molecules, through particles that are suspended in the atmosphere, and by the surface. This reflective power is called, a bit mysteriously for the uninitiated, albedo. This varies between zero and one and the higher it is, the more reflective the surface; an albedo of 0.9 (a level that new snow can reach) means that this surface reflects 90% of the incoming solar radiation. The net result of this albedo, whose level on a planetary scale is close to 0.3 (30%), is that the energy truly absorbed is 235 Wm^{-2}.

From one year to another, our planet holds a relatively constant temperature, which indicates that it is in thermodynamic equilibrium. In other words, the 235 Wm^{-2} of solar energy must be compensated for by an equivalent flux emitted toward space. In fact, every celestial body emits a radiation whose wavelength becomes smaller as its temperature rises. This is the case for the Sun and the Earth. But whereas the Sun, whose average surface temperature is close to 6,000°C, emits in the visible around 0.6 microns (μm) and in a range covering the ultraviolet up to the close infrared of 0.2 to 4 μm, the Earth emits in the infrared with a maximum intensity centered around 12 μm. If the atmosphere were perfectly transparent to infrared radiation, that radiation would be emitted by the Earth's surface, whose mean annual temperature would be −18°C. This would be the case if air were formed of only its three major components: nitrogen (78%), oxygen (21%), and argon and other rare gases (0.9%).

Fortunately this doesn't happen, thanks to a series of minor compounds. Formed of at least three molecules, they have a more complex structure than that of oxygen, nitrogen, and argon. It is thanks to them that infrared radiation emitted on the surface is absorbed by the atmosphere. The atmosphere reemits it into space but in a weaker quantity because the atmosphere is colder than the surface. The difference, around 150 Wm^{-2}, is sent back to the ground and serves to heat the lower layers of the atmosphere in which it is in some way trapped by the action of those minor compounds. Analogous to the gardener's greenhouse, those compounds are designated by the generic name of greenhouse gases. Water vapor (H_2O) is the primary greenhouse gas; it makes up 60% of the greenhouse effect. Next come carbon dioxide (CO_2) and ozone (O_3), which, respectively, represent 26% and 8% of the greenhouse effect. Methane (CH_4) and nitrous oxide (N_2O) account for the remaining 6%. The other contributors, some of which are pure products of human activity, such as the halogen compounds (CFC, etc.), encompass only around 0.2% of the overall greenhouse effect. However, they contribute to more than 10% of its current increase, which we call the additional greenhouse effect. Water vapor, carbon dioxide, ozone, methane, nitrous oxide, and so forth have been present in the atmosphere throughout time: this greenhouse effect is thus a completely natural occurrence. It is extremely beneficial, moreover, because it is hard to imagine how life could have developed in a mean temperature of −18°C. The temperature of +14°C reached thanks to this natural greenhouse effect is much more favorable.

How are the 235 Wm^{-2} that the Sun brings to the atmosphere used? A bit less than 30%, or 67 Wm^{-2}, are absorbed by the atmosphere and, for a large part, by the ozone in the ultraviolet, protecting us from that radiation and thereby making it possible for life outside the oceans to develop. This absorption takes place mainly in the stratosphere, the upper part of the atmosphere in which concentrations of ozone are highest and thus participate in the warming of that region; this is why the temperature in the troposphere (in the lower part of the atmosphere) decreases with altitude and increases in the stratosphere. But the greatest part of the Sun's radiation, 168 Wm^{-2}, traverses the atmosphere. About 100 Wm^{-2} is used to heat the surrounding air, of which most, 78 Wm^{-2}, enables the evaporation of water on the surface of oceans and continents, a process that plays a major role in the cycle of water and the redistribution of energy. In fact, this "latent heat" is used to evaporate water molecules and is restored to the atmosphere when those molecules are condensed, thereby redistributing energy depending on the formation of precipitations.

The redistribution of energy and the water cycle are at the heart of the functioning of the climatic machine. The energy budget that we have just presented is balanced on average over the entire planet, but it is not balanced locally. The reason for this is simple. On average, the Sun's radiation decreases annually by more than a factor of 2 between the equator and the poles, whereas the infrared radiation varies little (on the order of 10%). This isn't surprising that it is warmer at the equator. The excess is transported by the atmosphere and the ocean to the poles, which would be even colder without that contribution. About half of that contribution is carried out by the atmosphere, thanks to the winds and the water vapor that accompanies the air masses; the other half is brought by the ocean and its currents.

The Greenhouse Effect Due to Human Activity: A Slow Awareness

In the first two parts of this book we looked at the greenhouse effect through the work of Joseph Fourier and Svante Arrhenius in the nineteenth century. We might have added the Swiss Horace Bénédict de Saussure and his helio-thermometer, which he created to record the intensity of the solar flux at different altitudes, the French scientist Claude Pouillet, who demonstrated that thanks to the greenhouse effect the planet is warmer by several dozen degrees, or the Irish scientist John Tyndall, who explained that certain gases absorb infrared radiation and retain it within the atmosphere.

Then without saying anything—or almost nothing—about the twentieth century, we looked at Antarctica and the great ice core drillings of Vostok and Dôme C, which revealed that the great glaciations of the Quaternary were characterized by a close interaction between climate and greenhouse effect.

Arrhenius's calculations estimated a decrease of 40% in the concentration of CO_2 in a glacial period, an estimate which, somewhat by luck, proved to be correct. Those that were connected to a warming associated with a doubling in concentrations of carbon dioxide, estimated at around 5°C, appear a bit high in view of the current estimates, but they remain in the realm of the plausible. However, that prediction was largely ignored because Arrhenius foresaw that this doubling would only occur in the next 3,000 years. Granted, given the increase in emissions due to an increased use of carbon, he revised that number to a few centuries some dozen years later. But at the beginning of the twentieth century, climatologists were concerned more with explaining the glacial ages than with speculating on the future of our climate.

There was almost universal conviction at the time that the ocean or, by default, the vegetation would be able to absorb that excess carbon dioxide. However, just before World War II, the Englishman Guy Callendar deduced from the data available at the time that concentrations of CO_2 had already increased by 10%. He drew attention to the ocean's limited capacity for absorption, noting that the surface waters could rapidly become saturated. Callendar saw rightly, at least qualitatively. It was, however, necessary to wait until the 1950s for the idea that human activity was modifying the composition of the CO_2 in the atmosphere to be put forth, and thus the greenhouse effect took hold within the scientific community.

Three American researchers, Hans Suess, Roger Revelle, and Charles Keeling, played a major role. From an analysis of carbon 14 in tree rings, dated year by year, Suess showed that the concentration of that carbon isotope in the atmosphere had gradually decreased in the preceding years, which one would expect due to the emission of CO_2 coming from very old fossil fuel without carbon 14. Revelle was interested in the absorption of CO_2 by the oceans. Two years later, in 1957, Revelle and Suess explained why a part of the CO_2 remained in the atmosphere rather than being absorbed by the oceans. Persuaded of the need to measure very precisely the evolution of atmospheric CO_2, Keeling took the bull by the horns: in 1958 during the International Geophysical Year, he established the first station to measure the concentration of CO_2 near the summit of Mauna Loa in Hawaii.[1]

Others would follow, henceforth establishing a network of relatively dense stations, but they are still not enough. Let's first mention the precise measurements taken at South Pole Station. They began to be taken simultaneously with those of Mauna Loa and quickly confirmed the global nature of what was henceforth known as the Keeling curve, due to the position of the South Pole in the middle of Antarctica and thus beyond any significant local source of CO_2. Thanks to the initiative of Gérard Lambert, who, at the beginning of the 1980s, proposed building a station on the island of Amsterdam in the southern part of the Indian Ocean, France contributed significantly to this network through a team from the LSCE.

The results were not long in coming. Using monthly averages, Keeling revealed the existence of regular seasonal variations in CO_2 linked to the rhythm of development of vegetation. In the spring, vegetation grows while assimilating atmospheric CO_2, which consequently diminishes, then increases again in the winter; vegetation no longer plays its role as a CO_2 pump, and the decomposition of the dead leaves releases their carbon in the form of CO_2. These seasonal variations, which are very marked in the Northern Hemisphere, are much less pronounced in the Southern Hemisphere where the continents are much smaller. But regardless of the site, these variations are characterized by a huge increase in their mean annual levels and are practically identical from one place to another: 315 ppmv in 1958, 355 in 1992, and 387 in 2009 (Figure 11.2)—levels that the results from the ice cores of Vostok and Dôme C tell us have not been observed in at least 800,000 years.

This increase in concentrations of CO_2 has been shown in an increasingly detailed way for fifty years now. Carbon 13, another isotope of carbon, represents less than 0.1% of the carbon present on Earth (carbon 12, which amounts for more than 99.9% of the carbon on the Earth, is the main isotope of carbon). Analysis of it has shown that the CO_2 increase is indeed linked to fossil fuels. Compared to the CO_2 present in the oceans and in the volcanic or geothermal emissions, vegetation is indeed slightly lacking in carbon 13 because it has some difficulty using this heavy isotope during photosynthesis. The slight decrease in carbon 13 measured in the CO_2 of the atmosphere proves the contribution of fossil fuels, since these are of vegetal origin. In addition, the oxygen-nitrogen relationship decreases as the quantity of CO_2 increases, as we might expect when it comes from the combustion of oil, gas, and coal.

Faced with this increase, it is important to know how the concentration of CO_2 has evolved throughout the first part of the twentieth century, before 1958, and during the last centuries and millennia. Here, too, the polar ice proves to be extremely precious. This time, the objective is not to go back far in time; it is to have the most detailed records possible of the recent periods. To do this, we must use sites where there is a lot of accumulation. Situated in

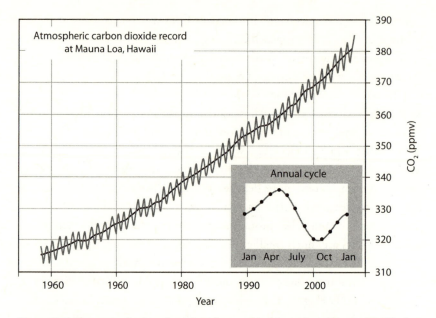

Figure 11.2. The atmospheric carbon dioxide record in Mauna Loa, Hawaii. The Keeling curve: variations in atmospheric concentrations of carbon dioxide shown in parts per million in volume (ppmv) measured at Mauna Loa. Seasonal variations are superimposed on the curve indicating the increase in the mean annual level, showing a typical annual cycle in the insert.

the coastal regions of Antarctica, they must also be cold enough: a mean average temperature below −20°C indeed avoids any summer melting of the ice that would modify its content of CO_2. Such sites have been identified: for example, at Law Dome, the high accumulation (70 centimeters per year) has enabled our Australian colleagues to use an overlap of more than twenty years between the record obtained from the air bubbles trapped in the ice and the direct measurements achieved through the atmospheric network. The correlation is excellent and provides proof of the viability of the measurements taken from the ice. The increase in CO_2 is well documented since the beginning of the industrial age. Before then, the concentrations were close to 280 ppmv and were even at 260 ppmv at around the beginning of the Holocene (around 10,000 years ago).

We had to wait until the 1970s for regular and precise measurements in

these observation networks, which were initially put into place to follow carbon dioxide, of other greenhouse gases (methane, nitrous oxide, and halogen compounds).

How Did We Get to This Point?

We must face the facts. For two hundred years our various activities have rapidly and greatly modified the composition of our atmosphere. Using the most recent IPCC report[2] and the Carbon Budget 2009 established by the Global Carbon Project,[3] one can describe how this happened.

Let's begin with carbon dioxide, the gas most responsible for an anthropogenic increase in the greenhouse effect. Each time we burn coal, gasoline, natural gas, or wood, we produce CO_2. This is also the case when we produce cement. Since 1750 the concentration of this gas in the atmosphere has gone from 280 to 390 ppmv (at the end of 2010) (see Figure 11.3). Its increase comes on the one hand from the modification of the land use, in particular from deforestation, and on the other from fossil fuels and cement plants. The latter contribution has gradually taken over from the former, which was dominant during the nineteenth century. At the end of 2007, the cumulative emissions linked to fossil fuels went beyond 330 billion tons of carbon (gtC). These emissions are constantly increasing; they have more than doubled since 1970. An exception: in 2009, they were the second highest in human history; they decreased from the previous year by 1.3% due to the global financial crisis, but in 2010 they increased again by 5.9%.[4] They are now close to 9.2 gtC/yr (8 in 2005; 8.7 in 2008; 8.4 in 2009), a bit less than 4% of which is due to cement factories and the rest to fossil fuels. The emissions linked to deforestation, which are difficult to calculate, are more than 1.5 gtC/yr and thus contribute less than 20% of the emissions that globally are close to 10 gtC/yr. Fortunately not everything remains in the atmosphere. Vegetation (despite the deforestation) and oceans absorb CO_2 roughly equally (each around 2 gtC/yr). The result: each year 45% of total CO_2 emissions remains in the atmosphere. The increase in the concentration of CO_2 has never been so high: 1.9 ppmv each year from 2000 to 2009.

Looking at the other greenhouse gases, we will distinguish between those that stay in the atmosphere for a long time (methane, nitrous oxide, halogen compounds) and those that react rapidly, such as ozone. Each of those mol-

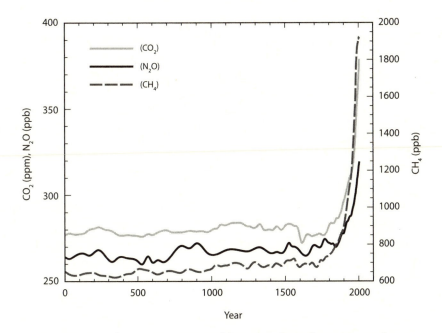

Figure 11.3. Atmospheric concentrations of the principal greenhouse gases in the past 2,000 years, showing the increase in those of human origin since the beginning of the industrial era, around 1750. Source: IPCC, *Climate Change 2007: Fourth Assessment Report* (Cambridge: Cambridge University Press, 2007).

ecules has a very different heating power: it is 23 times greater for a methane molecule than for a CO_2 molecule, and more than 10,000 times greater for certain halogen compounds. We compare these different compounds by measuring the quantity of additional energy available in the atmosphere due to the increase in their concentration since 1750, the year of reference: we call this "radiative forcing." Its increase thus corresponds to an increase in the greenhouse effect. That of CO_2 is estimated at 1.66 Wm^{-2}.

The second gas in importance that contributes to an increase in the greenhouse effect linked to human activity is methane, which is produced when organic matter decomposes in an anaerobic way, that is, outside the presence of oxygen. This occurs naturally in swamps and humid soils, sources to which are added termites, oceans, vegetation, and (we will come back to this) carbon hydrates. Human activities release CH_4 through the production of energy from natural gases and coal, the treatment of garbage and its burial, raising livestock, growing rice, and the combustion of biomass. Anthropogenic emissions represent close to 70% of the 600 million tons emitted, on average, every year. We should note that these emissions have nothing to do with those of CO_2; we speak in millions and not in billions of tons. Concentrations are also much weaker, more than 200 times less, or 1,774 ppbv (1.774 ppmv) in 2005. But they have greatly increased since 1750 since they have been multiplied by 2.5 (Figure 11.3), which corresponds to a radiative forcing of 0.48 Wm^{-2}. Let's add that once emitted, methane, which is eliminated through chemical oxidation, has a lifetime in the atmosphere of around ten years. This oxidation produces carbon monoxide (CO), CO_2, and water vapor. The oxidation of methane is, moreover, at the origin of the increase in the quantity of water observed in the stratosphere, an increase responsible for a radiative forcing estimated at 0.07 Wm^{-2}.

The question is often asked: do human activities contribute to an increase in the quantity of water vapor in the atmosphere? Their direct contribution is completely negligible because the recycling of water in the troposphere occurs over a very short amount of time, a few weeks at most. The quantity of water vapor is indeed increasing in the atmosphere but is doing so in reaction to the climate warming, which gradually affects the oceans, which consequently evaporate more.

The third contributor in importance to an increase in the greenhouse effect linked to human activities is the halogen compounds, which have contributed to a radiative forcing of 0.34 Wm^{-2}. The concentration of these constituents of purely anthropogenic origin, as is the case with the chlorofluorocarbons (CFC), has rapidly increased since World War II when their production and use became universal, whether as propellant in aerosols or as refrigerants. These compounds, which free chloride or bromide atoms into the upper atmosphere, are doubly toxic: they participate in the destruction of the ozone layer and are powerful greenhouse gases. Since the signing of the Montreal Protocol and its amendments, the production of the CFCs and of certain other compounds have been stopped. This has enabled a decrease in atmospheric concentrations of those compounds that have a short life and has slowed down the rate of increase for those that remain in the atmosphere a long time. By contrast, the substitute products, hydrochlorofluorocarbons

(HCFC) and hydrofluorocarbons (HFC), less dangerous vis-à-vis the ozone layer, are largely contributing to the increase in the greenhouse effect linked to human activities and their concentration is increasing, just like that of the perfluoride compounds (PFC), also a greenhouse gas.

Nitrous oxide (N_2O), whose natural sources are on the same order as those connected to human activities, has also increased by close to 18% since 1750 (Figure 11.3). The natural sources come from the ocean and from vegetation, while human activities contribute to emissions through the transformation of nitrogen fertilizers, the combustion of the biomass, raising livestock, and industrial activities such as nylon production. Once emitted, nitrous oxide remains in the atmosphere for a long time—more than 110 years—before being destroyed in the stratosphere. Its contribution to radiative forcing is 0.14 Wm^{-2}.

At this point it is interesting to evaluate the contributions to an increase in the greenhouse effect by large sector and by country, once the contribution of each gas has been taken into account. Thus in 2004 on a global level the production of energy contributed to 26% of the increase; industry to 19%; land use change and deforestation to 17%; agriculture to 14%; transportation to 13%; residential, commercial, and tertiary sectors to 8%; and waste to 3%. In that same year the situation was very different in France. The greatest contribution to the greenhouse effect there, 26%, was linked to transportation. It was followed by industry at 20%, then in equal parts at 19% by the residential, commercial, and tertiary sectors and those of agriculture and forestry; the production of energy and waste contributed, respectively, 13% and 3%.

These assessments have enabled us to identify the contributions of each country—and this is a crucial point in international discussions—which has revealed enormous disparities per capita among developed and developing countries. Let's look at the CO_2 emissions due to fossil fuels for 2005. The United States, with 21% of total emissions, and China, with 19%, were the largest contributors (China has now surpassed the United States). But the scene is different when we use these figures to calculate each country's per capita carbon emission; in the United States this is nearly 5.5 tons of carbon per inhabitant, five times more than in China (1.11). But emissions do not exceed a few hundred kilograms per inhabitant in some Asian countries (such as Vietnam) and in Central America (such as El Salvador) and are only a few dozen kilograms in Haiti and in many African countries; 100 times less than that of an American. A French person emits three times less than does an American. With 1.8 tons of carbon emitted per inhabitant, a French person emits less, by around a third, than a German or an Englishman because close to 90% of France's electricity comes from sources that do not produce CO_2—a bit less than 80% is nuclear, around 10% is hydraulic, and less than 1% is wind.

Finally, ozone (to which we will return in chapter 17) is also a climatic gas. It enables the maintenance of life on Earth by preventing the ultraviolet solar rays with short wavelengths from reaching the surface, but it also absorbs the Earth's infrared radiation and in so doing participates in the greenhouse effect and in its variation. It does so in two ways. First, the decrease of ozone in the stratosphere decreases the corresponding greenhouse effect by around 0.05 Wm^{-2}. Second, various gaseous species contribute to the production of ozone in the troposphere; this is the case of the CO already cited, nitrogen oxides (NO_x, not to be confused with N_2O), methane, and other hydrocarbons. These are called "indirect greenhouse gases," the production of which is linked, in one way or another, to human activities. As a result, since the preindustrial period there has been an estimated increase of 30% of ozone in the troposphere with a radiative forcing of 0.35 Wm^{-2}.

It is thus incontestable: through our activities we have been rapidly and greatly modifying the composition of our atmosphere for two hundred years. What is more, we know precisely how the greenhouse effect has increased as a result of human activity, let's say by about 10%. The assessment is simple: the observed variations in concentrations of carbon dioxide, methane, halogen compounds, nitrogen protoxide, and ozone have increased the greenhouse effect by around 3 Wm^{-2}, approximately 55% of which is linked to an increase in CO_2—an increase of more than 1% compared to the 235 Wm^{-2} available to heat the lower layers of the atmosphere and of which a large part results from human activity post–World War II. The questions then become: Is the climate in the process of warming? If so, are we responsible for this?

CHAPTER 12

Have Humans Already Changed the Climate?

In 1896 Svante Arrhenius brought attention to the fact that humans were in the process of changing the amount of carbon dioxide in the atmosphere and that as a consequence our planet would warm up by 5°C from that point to the end of the twentieth century, according to his estimates. It was only eighty years later that this risk of warming and its potential consequences were taken seriously. This awareness quickly led to the establishment of the Intergovernmental Panel on Climate Change (IPCC) in 1988, which in 2007 concluded that our activities are very likely at the origin of a marked warming that we have been experiencing since the mid-twentieth century .

The Time of the Pioneers

Until the 1970s very few research teams were looking at the impact of human activity on the climate. It is true that after an initial period of warming up to the beginning of the 1940s, temperatures stabilized, then slightly decreased. Moreover, in 1975 Willi Dansgaard,[1] the Danish glaciologist, a pioneer in the reconstruction of past climates from the ice of Greenland, extrapolated one of the climatic records he had obtained at Camp Century northwest of that ice cap and predicted a future cooling of the planet. The idea at that time was that the length of the interglacial period in which we are living, the Holocene, would end quickly. That fear was in fact unfounded since the Earth's orbit is such that a return to a glacial period is not anticipated for another 10,000 years or even more. But thirty years ago, we were far from the cry of alarm that we are sounding today regarding the effect of human activity on the climate.

A few pioneers did, however, attempt to have their voices heard. In the 1930s Guy Callendar, who had correctly seen an increase in CO_2 in the atmosphere, sought to establish a connection between the warming that had been occurring since the beginning of the century and an increase in the greenhouse effect that resulted from an increase in CO_2. With this idea, he established a curve showing the average temperature over all the continents from series of temperatures measured in around two hundred meteorological stations. Wladimir Köppen, already cited for his work in paleoclimatology, had undertaken the same study at the end of the nineteenth century, but Callendar had much longer records that enabled him to show and to correctly document the warming that had begun since the beginning of the twentieth century. He postulated that this warming might be a result of the use of coal and estimated that the average warming linked to a doubling of CO_2 was 2°C, with more elevated levels in polar regions. Like Arrhenius, he saw only beneficial effects in this warming that would make northern climates milder and would delay a return to a glacial period.

The Awareness

Modelers were really the first to become aware of the potential importance of the influence of human activity on the climate. The postwar period saw the appearance of the first calculators, which at the time were quite rudimentary. Forecasting the weather had been within the realms explored by the pioneers of those machines that would revolutionize our world and, in particular, that of scientific research. These models rested on a system of physical equations that enabled scientists to describe movements of the atmosphere and the water cycle from evaporation on the surface of the ocean to the formation of precipitation. These equations also took into account exchanges of energy— that provided by the Sun, a part of which is used during evaporation, and that freed when the water vapor is condensed in the atmosphere and forms the fuel of the climatic machine. The first applications were in the realm of meteorology, which involves the prediction of weather disturbances whose individual evolution we can follow only over a few days; beyond a dozen days the atmosphere loses its memory.

Climatologists are interested in longer time scales—a month and beyond—and their forecasts bear only on the average temperature values, quantities of precipitations, or characteristics of the atmospheric circulation, not on following specific disturbances. We are all well aware of the progress made in forecasting the weather, progress that went hand in hand with the increase in the capacity of the means of calculation, but also of those dedicated to observation. The climatic models were also much improved even if, as we will see, much uncertainty remains. They are, we emphasize, based on equations that are identical to those of meteorological models.

The first experiments, carried out in the 1960s, enabled us to verify that the models could simulate the great characteristics of the climate: mean annual levels of temperatures and precipitations, winds, seasonal cycles, and geographic distribution of climatic zones. But climatologists quickly became interested in analyzing the ability of these models to account for climates different from the one in which we are currently living. In principle, it was easy for a climate modeler to modify the composition of the atmosphere and to make a new simulation even if the number of simulations was at the time fairly limited due to the lack of computing power. The first experiments of this type, carried out in the 1970s, looked at the effect of a doubling of the quantity of carbon dioxide compared to its current value.

Granted, the results differed noticeably from one model to another; thus the warming predicted by four different models (three American models, one English) in the event of instantaneous doubling of the concentration of carbon dioxide varied from 1.5 to 4.5°C. This factor of three, in the value of what we call "climate sensitivity," results for the most part from the way in which the formation of clouds was dealt with. Their optical properties cause them to both absorb and reflect solar radiation. They are, moreover, affected in different ways depending on whether they are "high clouds" or "low clouds." This complexity means that the behavior of cloud systems is difficult to account for in models. Furthermore, two versions of the same model (that of the UK Meteorological Office), between which only this taking into account of clouds has been modified, forecasted respective warming of 1.9 and 5.2°C.[2]

But beyond these uncertainties, all the models (this assertion remains nearly verified now that the number of simulations has been multiplied by more than a factor of ten) forecasted a warming. What is more, this warming is systematically greater than that which would be obtained in the absence of any climatic feedback. This would be the case if our atmosphere were composed of only nitrogen, oxygen, and argon, without any of the greenhouse gases we have mentioned—water vapor, carbon dioxide, methane—and if the albedo of our planet were fixed once and for all. We would then expect a warming of 1.2°C for a CO_2 doubling, less than the lowest level of 1.5°C predicted by the models.

Some amplifying mechanisms thus dominate the climate's response during an increase in the greenhouse effect, and this can be proven. Thus a warming of the atmosphere will, with a certain delay, be transmitted to the layers of the ocean surface. This brings an increase in evaporation, which grows exponentially as a function of the temperature, and consequently of the quantity of water vapor present in the atmosphere. Since water vapor is itself a greenhouse gas, the radiative forcing is amplified. The decrease of sea ice and snow-covered surfaces, which are very reflective vis-à-vis solar radiation, in response to this gradual heating of the oceans constitutes a second amplifying factor because it is then replaced by much more absorbent ocean and continental surfaces.

A large majority of the scientific community has been convinced of the amplitude of the problem and of the necessity of analyzing all its facets. The first results obtained from the Antarctic ice at Vostok also suggest that the climate is an amplifying system. Those results have thus helped raise awareness of the gradual warming of the planet. And that awareness has been further amplified by the fact that the average temperatures measured on the surface of the Earth in the 1980s are clearly rising.

Some scientists believe it is necessary (and important) to make people aware of warming through the media. In the United States, Jim Hansen directs a climate research institute in New York that is part of NASA. In the spring and summer of 1988, a drought that had the potential to be catastrophic affected a large part of North America. Hansen suggested that it was the result of climate warming and that the greenhouse effect had already arrived. He placed a wager—which he won—that warming would increase and that the hottest year up to that point would be surpassed within the following three years. We know Jim Hansen well, as we have collaborated with him. He is a first-rate scientist, and even if some criticize him for being quick to react, he did so with the necessary caution and all scientific honesty. Many others are sounding the alarm. From the closed circle of climate specialists, the issue of the greenhouse effect became known by the general public by the end of the 1980s.

The Establishment of the IPCC

The results of the climate modelers were taken seriously and motivated expert scientists to hold meetings and write reports on climate, one of which, published in 1979 at the initiative of the NSF, had very well-documented conclusions. The World Meteorological Organization (WMO) initiated the World Climate Research Programme (WCRP) and organized the first World Conference on Climate. It concluded that human activity was likely to influence our climate in a significant way. But it was during the following decade that true awareness grew not only within the scientific community but well beyond it as people began to realize the scope of the consequences of this climatic change, both ecologically and economically. In 1985 the Villach conference in Austria on the evaluation of the role of carbon dioxide and other greenhouse gases marked a turning point; the consensus that arose among scientists at the meeting was that as a result of an increase in concentrations of greenhouse gases one could expect, during the first half of the twenty-first century, an increase in the average temperature greater than what had ever been known before.

The scientific community quickly got organized through both the WCRP, which developed rapidly under the direction of the French climatologist Pierre Morel, and the International Council for Science (ICSU), which brought together scientific organizations from around the world. Aware of the strong interactions that exist between the climate and the environment, in 1986 the ICSU launched an ambitious program dedicated to the study of the geosphere and the biosphere: the International Geosphere-Biosphere Program (IGBP). The WCRP and the IGBP were two of the four components of what has henceforth been known as the Global Change program. (The two others are devoted to biodiversity and the human dimension of climate change.) Thanks to these international initiatives, which are now well established by research organizations in many countries and in Europe, an understanding of the complex mechanisms that govern the evolution of the climate has largely been elevated during the last twenty years.

Governments did not delay in getting involved, aware that this was an issue that could not be ignored, especially since the protocol that prohibited the production of components contributing to the destruction of the ozone layer, signed in Montreal in 1987, had already established what could be undertaken on behalf of the global environment. With some distance, the decisions made within that framework appeared completely right and, thanks to them, we can hope for a slowing of the decrease in the ozone layer in the decades to come, then a reversal of the process, and in the second half of the twenty-first century a return to the conditions that prevailed before human activity disturbed the ozone layer. The scientific consensus that was quickly established, the well-identified consequences, and the restricted number of producers largely facilitated the signing and then the implementation of the protocol. Although the issue of climate change presents many similarities to that of stratospheric ozone, the former is much more complex. The existence of a connection between human activity and climate warming then rests essentially on the predictions of models, which we know are still rudimentary, and the consequences of this warming are not well understood. Furthermore, if measures to limit the use of fossil fuels must be created and implemented, it will be difficult for the general public to accept them since economic development and personal comfort are intimately mixed with consumption of energy. The first step thus consists of establishing a diagnosis.

It was with this objective that in 1988, under the joint auspices of the United Nations Environment Programme (UNEP) and the WMO, the IPCC was created. From the beginning the IPCC was interested in assessing the available knowledge in three distinct ways: looking at the scientific aspects of the evolution of the climate (Group I); examining its impacts and analyzing the measures of adaptation (Group II); and assessing the options for mitigating climate change (Group III). Four assessment reports have been published (1990, 1996, 2001, and 2007). Each report is divided into chapters written by a team of a dozen authors from different countries. To produce each chapter the authors solicited contributors working in a specific field. Summaries around fifty pages long of each of the four reports (close to a thousand pages) were written, followed by "summaries for policymakers," which were much shorter and written in a more accessible style. Everything is brought together by a synthesis report. Once written, each report was commented upon by the scientific community (reviewers) and then by representatives of governmental as well as nongovernmental organizations. It takes more than two years for a text to be written and reviewed, approved by the scientific community, and then proposed to governments. Commentaries from different sources (the scientific community, governmental agencies, nongovernmental organizations) are taken into consideration by the editors, and the texts are then edited. If the editors consider a commentary unfounded, which is sometimes the case, they have to justify the reason or reasons that led them not to take it into consideration. After the publication of the third report, editors were hired to ensure that all of the commentaries are taken into account. As an example, for Group I alone there were 160 principal authors who participated in the fourth report and nearly 600 scientists including editors, contributors, and reviewers.

Then comes the last step before publication: approval by the governments who are members of the IPCC (more than a hundred countries). The summaries for policymakers were discussed line by line by delegates from member countries and approved, after possible changes, during meetings that representatives from nongovernmental organizations can attend as observers. The general rule is to reach consensus, which is sometimes very difficult despite the efforts of the IPCC Group I co-chairs and the authors present at the final meeting. They try to reach a solution that is acceptable to everyone and reflects the complete reports. The content of the reports is not discussed again, but that of the extended summaries is also subject to approval, and the coherence between the different stages of the reports receives a great deal of attention.

The rest of this chapter and the next chapter look mainly at the work of Group I, which deals with the science of climate change and in particular with the physical aspects of those changes. In the first IPCC report, the role of human activity in the modification of the composition of the atmosphere and the associated increase in the greenhouse effect were fully recognized. The rise in temperature and other climatic parameters was, in each of the successive reports, scrutinized. Two questions came up repeatedly: Has human activity already changed the climate? This question is examined in the present chapter. The second is: What climate for the future? It is discussed in the following chapter.

It is entirely legitimate to wonder whether the warming that we are experiencing on a global scale has a connection with the known increase in the greenhouse effect. In 1990 that warming, estimated at 0.3–0.6°C over the last century, was just getting under way. Without mentioning the glacial periods, we need only turn to the recent past to observe that there is no need for man to intervene for the climate to change in such a notable way, as it did during the twentieth century. For the period that we call the Little Ice Age (from the mid-fifteenth century to the late nineteenth century), this is beyond question. It is supported by much evidence, such as the advance of Alpine glaciers and Flemish painters' renderings of rivers covered with ice at that time. During that cold period, which reached its height between 1550 and 1770, the temperature was, at least in western Europe, at a minimum one degree colder than it was in the twentieth century. By contrast, the beginning of the last millennium, a period in which the south of Greenland was possibly a land a bit more hospitable than it is today, was relatively warmer. It is difficult in these conditions to assert that this early warming was already connected to human activity.

The first report (1990) of the IPCC responded to the question regarding human activity's impact on the climate in the following way: "The size of the warming over the last century is broadly consistent with the prediction by climate models, but is also of the same magnitude as natural climate variability. If the sole cause of the observed warming were the human-made greenhouse effect, then the implied climate sensitivity would be near the lower end of the range inferred from models. Thus the observed increase could be largely due to this natural variability: alternatively this variability and other human factors could have offset a still larger human-induced greenhouse warming."

Scientists noted that the greenhouse effect was increasing and that the climate was warming, but at the time they were not at all in a position to establish a relationship of cause and effect.

The Problem of Aerosols

If we look at the increase in greenhouse gases since the beginning of the industrial era, the models of the evolution of the climate predicted a warming of nearly 1°C, roughly double of that observed. But those predictions did not take into account all the aspects connected to human activity. Thus they ignored the effect of cooling resulting from the presence of microscopic particles, solids or liquids, suspended in the air, which are called aerosols. This was one of the advances of the second report of the IPCC (1996), which focused on them, even though the exact role of those aerosols on climate was at the time and still is not perfectly understood.

Aerosols can have different origins: they can be natural (desert, marine, or volcanic), anthropogenic, connected to the use of fossil fuels, forest fires, or pollution in industrial regions. Their lifetime is very short, on the order of a week, and the geographical distribution of their sources is very heterogeneous. They diffuse and absorb solar radiation and, to a lesser degree, infrared radiation. In general they have a cooling effect, except for carbon soot and for the aerosols issued from biomass fires. From recent measurements, including data obtained from satellites, we can estimate the combined radiative forcing of all of these aerosols at -0.5 Wm^{-2} with, however, an uncertainty as large as ± 0.4 Wm^{-2}.

This is called the direct radiative forcing of aerosols. But things are even more complicated because these aerosols also act on the microphysical properties inside clouds, some of which play the role of condensation nuclei. Through that indirect effect, aerosols modify the albedo of clouds and thus the radiative forcing associated with them. Once again, there is a cooling effect, of -0.7 Wm^{-2}, but with an even greater uncertainty in the estimates, since those range from -0.3 to -1.8 Wm^{-2}. So the effect of all human activity that involves greenhouse gases, the direct and indirect effect of aerosols, as well as other more marginal contributions, such as those linked to persistent contrails from airplanes, changes in vegetation, and in the deposition of aerosols on snow-covered surfaces, was still not known with great precision in the 2007 report. The effect of such activity was unquestionably dominated by the greenhouse effect, with a positive value of 1.6 Wm^{-2}, but, following the real effect of aerosols, the estimates were between 0.6 and 2.4 Wm^{-2}.

In retrospect, the authors of the second report could have been quite right since the direct radiative forcing was then estimated at -0.5 Wm^{-2} with an indirect forcing of comparable value. It was in fact sufficient to counteract part of the warming due to an increase in the greenhouse effect and to eliminate the discord between predictions and observations. More convincing for specialists was the inclusion of an entire series of indications corresponding to a series of imprints suggesting that the warming observed was likely not simply of natural origin. These indications rested on geographical, seasonal, and latitudinal comparisons of warming, whose predicted and observed

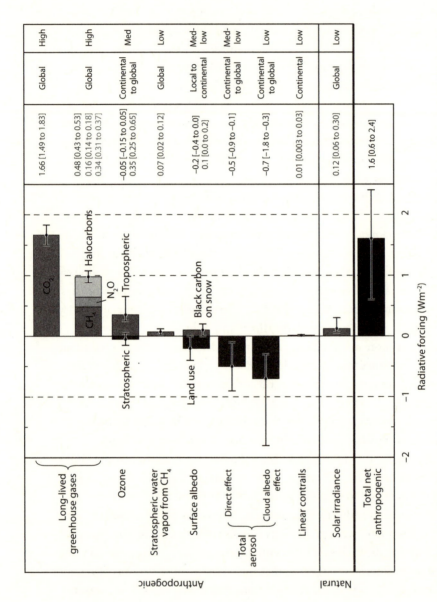

Figure 12.1. Variation between the beginning of the industrial era, around 1750, and 2005, in the principal factors of radiative forcing expressed in watts per square meter (Wm^{-2}). Source: IPCC, *Climate Change 2007: Fourth Assessment Report* (Cambridge: Cambridge University Press, 2007).

characteristics agreed all the more in that the models took into account the role of the greenhouse effect and of that of aerosols and not only the natural causes of climatic variability such as volcanic eruptions, which could provoke a notable cooling but of short duration, or weak fluctuations in solar activity.

None of these elements taken individually constituted proof, but their convergence in 1996 led scientists to conclude that despite uncertainties, "the balance of evidence suggests that there is a discernible human influence on global climate."[3] To suggest, even cautiously, that human activity was beginning to have an influence on the climate placed climate change at the center of the problems that our society would have to confront in matters concerning the environment and gave it an undeniable socioeconomic dimension.

The Climate in the Last Millennium

The assessment was refined between the 1996 and 2001 reports thanks in particular to a better knowledge of the variations in the climate during the last centuries. We quickly understood that it was absolutely crucial to distinguish between the warming of natural origin and that which might already be attributed to an increase in the greenhouse effect linked to human activities. There were many rather homogeneous sources of information for reconstituting that so-called recent climate. We are indebted to Michael Mann and his American colleagues[4] for having ingeniously integrated the data drawn from different natural archives—tree rings, coral, polar ice—with historical archives to create a curve describing the evolution of the mean temperature since the year 1000 for the Northern Hemisphere, where climatic series are more numerous. That mean temperature decreased slightly from 1000 AD to the end of the nineteenth century: it climbed to a medieval high point between the eleventh and fourteenth centuries and then sank during the Little Ice Age between the fifteenth and the nineteenth centuries. The associated variations in temperatures are weak, however, and the different regions of the Northern Hemisphere were in fact not affected in the same way or at the same time. This curve (Figure 12.2), called the "hockey stick" because of its overall shape, led the IPCC to conclude in 2001 that the warming of the Northern Hemisphere in the twentieth century is likely to have

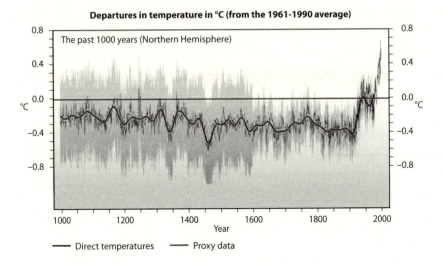

Departures in temperature in °C (from the 1961-1990 average)

Figure 12.2. Reconstruction of the temperature of the Northern Hemisphere during the last millennium (known as the "hockey stick" curve) from records derived from the analysis of tree rings, coral, ice cores, and historical archives (thin line), with a smoothed curve in a thick line. Source: IPCC, *Climate Change 2001: Third Assessment Report* (Cambridge: Cambridge University Press, 2007).

been the largest of any century during the past thousand years. It has since been the object of much criticism, to which the 2007 report mainly responded. That report indicated a greater variability in the temperatures of the Northern Hemisphere than the 2001 report did, with, in particular, colder periods in the twelfth to fourteenth centuries and in the seventeenth to nineteenth centuries. It concluded that average Northern Hemisphere temperatures during the second half of the twentieth century were *very likely* higher than during any other fifty-year period in the last five hundred years and *likely* the highest in at least the past 1,300 years. That conclusion left little doubt: recent warming is beyond natural variability.

To be able to attribute that warming to human activity we had to separate warming that was the result presumably of that activity from warming that naturally makes the climate vary, primarily volcanoes and solar activity. Volcanic eruptions and variations in solar activity are well documented for the twentieth century, climatic models have made important progress, and the computing power has progressed a great deal, allowing for an increase in the number of simulations. The conclusions are clear. Lengthy simulations, done while separately taking into account the natural and anthropogenic forcings, then by combining them, show that the warming of the last hundred years is most likely not due to natural causes alone. In particular, the noticeable warming of the last fifty years can only be explained if one takes into account the increase in the greenhouse effect. Thus this conclusion in the 2001 report: "Most of the observed increase in global average temperatures since the mid-20th century is likely due to the observed increase in anthropogenic greenhouse gas concentrations."[5] From "perhaps" in 1996 we are now at "probably" in 2001, which in our jargon corresponds to more than two chances out of three.

Warming Is a Certainty

Saturday, February 2, 2007: the IPCC Group I report dealing with the scientific elements of climate change had just been approved in Paris. Susan Solomon from the United States, co-president of that group with Qin-Dahe from China, presented the principal results before more than four hundred journalists from around the world. A sentence caught our attention because of its simplicity: "Climate warming is unequivocal."[6] At the end of the 1980s, as we have mentioned, warming was just beginning following a phase of light cooling, which, in the 1950s and 1960s, had followed the warming observed in the first half of the twentieth century.

Since then warming has accelerated. To be convinced of this one need only look at the data that enable us to follow the evolution of the mean temperature of the planet since 1850 (Figure 12.3). Warming has been more than twice as rapid in the last twenty-five years (close to 0.2°C per decade) than during the last hundred years (warming of 0.74°C over the period 1906–2005). Every recent year has been a warm one. With the exception of 1996, eleven of the twelve years in the period 1995–2006 have been warmer than all those that preceded them. And 2007 has just been added to that list. The year 1998 was particularly warm because of an important El Niño phenomenon, which was at the origin of a marked warming over a part of the Pacific Ocean, whereas 2007 was characterized by an inverse event, La Niña, without which it would have probably been a record year. It is a good bet, as Jim

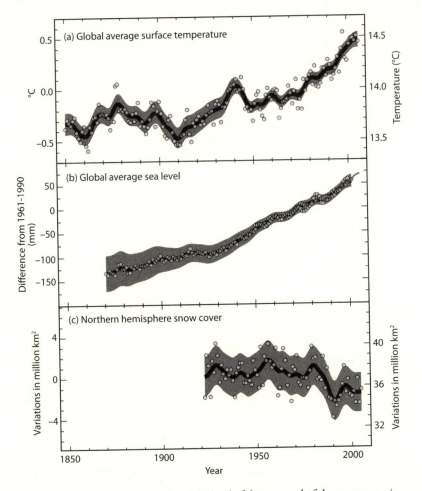

Figure 12.3. Change in temperature, of the level of the seas, and of the snow cover in the Northern Hemisphere. Source: IPCC, *Climate Change 2007: Fourth Assessment Report* (Cambridge: Cambridge University Press, 2007).

Hansen predicted, moreover, that the record of 1998 will be broken before 2010–11 at the latest, which has since been verified, although 2010 is only slightly warmer than 1998.

What is true for the planet is true for France. The national weather service, as we have seen, recently published a study in which the data available for France were examined closely. Once processed, they were used to establish maps of the variation in temperature during the twentieth century

(1901–2000). These revealed that the southwestern quarter, north of the Massif Central and part of the Alps, have warmed by more than 1°C. Everywhere else the increase is between 0.8 and 1°C except for a part of the northeastern quarter for which it is between 0.6 and 0.8°C. Except in those regions, warming is greater than that observed, on average, on the planet. And it has been particularly noticeable since 1985.

Beyond mean annual temperatures, many other indicators have led the IPCC to its precise diagnosis of an unambiguous warming. It is also visible in the oceans, whose average temperatures have increased from the surface to depths of at least 3,000 meters. The result is that oceans are expanding. This thermal expansion of the oceans, together with the melting of glaciers, glacial ice caps, and inland ice sheets, has played a role in the recent acceleration of the rise in sea level; the level of 3.1 millimeters per year observed over the period 1993–2003 is almost double what it was in the last century, estimated at 1.7 millimeters per year, or a rise of 17 centimeters since the beginning of the twentieth century. Another observation that testifies to warming, which we will return to, is a generalized melting of snow and ice. It is also interesting to note that the increase in the average quantity of water vapor in the atmosphere is compatible with the warming observed.

The fact that warming has clearly continued since the beginning of the twenty-first century is one of the elements that, in 2007, enabled us to strengthen our assessment vis-à-vis the role of human activity. Also, new data have shown that the variations of the solar activity since 1750 were previously overestimated by a factor of two. In addition, the effort made by the modelers of comparing simulations that take into account natural forcings on the one hand and all forcings on the other has been intensified. That exercise demonstrated that the influence of human activity was visible not only on the scale of average temperatures of the planet (Figure 12.4) but also on each of the continents taken individually, except Antarctica. It was thus with solid proof that the IPCC reported that "most of the observed increase since the mid-20th century is very likely due to the observed increase in anthropogenic greenhouse gas concentrations."[7] This was almost a certainty. The IPCC added that it is *likely* that increases in greenhouse gas concentrations alone would have caused more warming than observed because volcanic and anthropogenic aerosols have offset some warming that would otherwise have taken place.

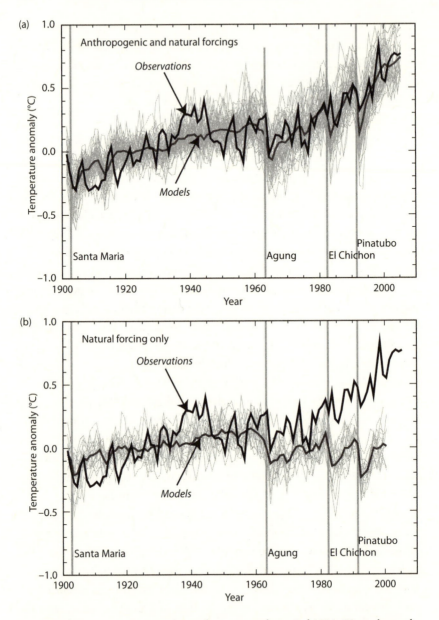

Figure 12.4. Variation in temperature with respect to the period 1901–50, as observed (thick line), and as predicted by a group of models taking into account both anthropogenic and natural forcings (a) and natural forcings alone (b). All the results of the models are represented by the gray area and their average value by the continuous, thinner curve. Source: IPCC, *Climate Change 2007: Fourth Assessment Report* (Cambridge: Cambridge University Press, 2007).

The Arguments of Skeptics

The scientific community have been presenting solid arguments for quite some time, but they are still the object of skepticism, in particular among various researchers from other disciplines and outside the scientific realm, whether in the world of political policymakers, large corporations, or the man on the street. In many cases, this skepticism has amounted to denying the impact of human activity on the climate or at least minimizing it.

This seems to be the appropriate moment to look at the steps taken by skeptics from various domains. Let us stress that we believe deeply in the virtue of scientific debate, a debate that is all the richer in that some arguments might provide contradictions to what gradually is becoming the dominant belief. Still, the process undertaken by skeptics must show as much scientific rigor in disproving the link between human activity and global warming as has the scientific community in linking the two. In any event, it is up to those defending the dominant point of view to convince the others. We fully agree with the conclusions of the IPCC, to which two of us, Jean Jouzel and Dominique Raynaud, have contributed. We also have a responsibility to respond to the arguments put forward by skeptics, among whom in France we encounter some in *Ma vérité sur la planète* (2007) by Claude Allègre. In 2010 Allègre, a geochemist, the recipient of the prestigious Crafoord Prize in 1986, and a former minister from 1997 to 2000, attacked what he called "the climatic imposture."[8] His colleague, Vincent Courtillot,[9] challenged the validity of certain data used in the IPCC reports. In France, as in other countries, their arguments resonate both with the general public and with some policymakers in the political and economic spheres.

The rise in climate skepticism has increased since 2009. With the partial failure of Copenhagen—we will come back to this—there was a campaign to discredit the way the IPCC functioned, as well as the conclusions published by that group of experts. That campaign began right before Copenhagen with "Climategate," which entailed the publication of a series of e-mails hacked from the site of English researchers at the University of East Anglia, including Phil Jones, who is well-known for his work on recent climate data. The campaign was strengthened by the revelation of errors identified in the 2007 IPCC report and was pursued in France and in other countries through the publication of various materials on climate skepticism, such as those we have just cited.

We won't dwell on Climategate, which among other things illustrates, through exchanges that are rather coarse, the healthy divergence of opinion that can exist among climatologists. The results of a parliamentary inquiry exonerated the University of East Anglia researchers who were accused of falsifying data by certain climate skeptics. In any event, we remain absolutely convinced that those e-mail exchanges, which a priori should not have gone beyond the private realm of the recipients, do not challenge the conclusions of the IPCC.

The "errors" of the IPCC are really regrettable and do it a disservice. They do, however, appear to be inconsequential. At the beginning of 2010, there was in fact only one established error. It concerned the assertion according to which the Himalayan glaciers will have lost 80% of their surface area by 2035. This could have been a simple typo (2035 might have been typed instead of 2350) or have been connected to the use of "gray" documents that had not been peer reviewed, but the origin of this error, recognized by the IPCC, has not been completely clarified. In any case it does not constitute a key point in the conclusions of a nearly three-thousand-page report. Another "error" was the IPCC's conclusion that 55% of the Netherlands was located below sea level. The correct number is 26%; an additional 29% could be affected by flooding. Thus it should have been noted that for 55% of the territory, there is a risk of flooding. As for other so-called identified errors (a decrease in harvests in Africa, an overestimation of the cost of climate disasters, a response by the Amazon forest to climate warming), skeptics have pointed these out in an effort to challenge certain conclusions, to which clear responses have been made by scientists directly involved.

Beyond these "errors," the very functioning of the IPCC has been questioned. In this context, it is interesting to note that the IPCC requested an external audit to examine on its functioning and the various procedures it has followed. With the assurance of the InterAcademy Council, the report disclosed at the end of the summer of 2010 found that the IPCC assessment process has been successful overall but also made recommendations that might be put into effect for the next report. These recommendations aim, in particular, to strengthen the IPCC's governance and the management (through the election of an executive director and the establishment of an executive committee) as well as the review procedures for creating the IPCC reports (with a more important role for the review editors). Other recommendations deal with improving the evaluation of the uncertainties behind the IPCC assessments, as well as communication and transparency.

Of more concern is the rise in skepticism widely conveyed by the media. It is a skepticism that in itself is legitimate but which we believe is not based on true scientific arguments. It is difficult to present an exhaustive inventory of the arguments here—arguments made by the skeptics—but we will attempt to illustrate our position by examining those that are most often put forth to place in doubt the key conclusions of the 2007 IPCC report. We will focus on a few key points, but the interested reader should refer to the RealClimate website, which is sponsored by American scientists such as Gavin Schmidt and Ray Pierrehumbert and presents an in-depth study of the debate. In France, journalists such as Stéphane Foucart at *Le Monde* and Sylvestre Huet at *Libération*, who is also the author of *L'imposteur c'est lui*,[10] have revealed errors and manipulations published in the aforementioned work by Claude Allègre.

In the face of false accusations, an open letter, "Scientific Ethics and Climate Science," which we signed in April 2010 along with six hundred French scientists involved in climate research, demonstrates the scientific integrity of our research community and solicits the beginning of a truly professional and serious debate. It is that same notion of scientific integrity that more than two hundred members of the American Academy of Sciences put forward in a letter titled "Climate Change and the Integrity of Science," which was published in the journal *Science* in May 2010.[11]

As an example of results that have been distorted by climate skeptics, we cite the observation, used by Claude Allègre, according to which "variations in temperature preceded that of CO_2 by 800 years."[12] In fact the presentation of those results, of which one of us, Jean Jouzel, is coauthor, is skewed and biased. With Nicolas Caillon and our glaciologist colleagues, we wrote that the increase in the temperature *in Antarctica* preceded the variations in CO_2 by 800 years but also that these began a few thousand years before the melting of enormous quantities of ice accumulated on the continents of the Northern Hemisphere (Figure 12.5).[13] The difference is essential because that sequence is completely compatible with a contribution of an increase in CO_2 to the deglaciation of the large ice sheets of the Northern Hemisphere. And it takes a bit of bad faith to use those results with a view toward minimizing the role of human activity on the climate because, in any event—we stress

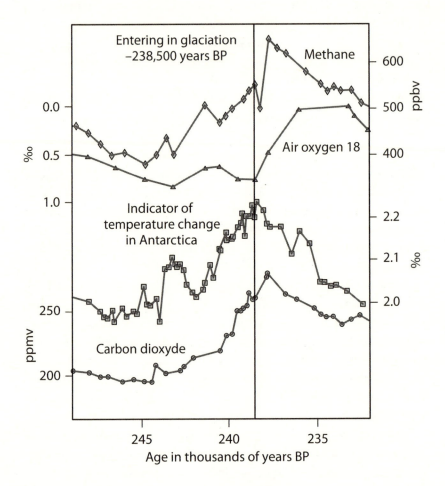

Figure 12.5. Sequence of events during a deglaciation, around 250,000 years ago. The vertical line corresponds to the beginning of the deglaciation in the Northern Hemisphere, around 238,000 years ago, as it is indicated by a modification in the content of oxygen 18 in the air and in the concentration of methane. Several thousand years before the deglaciation, the concentration of carbon dioxide began to increase, and then about 800 years later the temperature in Antarctica began to increase; around 238,500 years ago the temperature in Antarctica had already reached its maximum, and that would be true a few hundred years later for carbon dioxide. Source: Adapted from Nicolas Caillon et al., "Timing of Atmospheric CO$_2$ and Antarctic Temperature Changes across Termination-III," *Science* 299 (2003): 1728–31.

this here—the recent rise in the greenhouse effect has a different character. As a result of human activity (no one denies this), the rise in CO_2 preceded the global warming currently under way.

The ability to precisely measure an average global temperature, such as the curve produced by Phil Jones, one of those used by the IPCC, is itself doubted by skeptics. This curve, of course, was established with the greatest care, which involved correcting certain effects such as those linked to urbanization, the moving of certain stations, or local modifications in the environment. Recent studies have confirmed that both urbanization and changes in land use have had a negligible effect, less than six-thousandths of a degree per decade, on global temperature measurements. The IPCC does not rely on Phil Jones's curve alone but also on those produced independently by two American teams; the three series are completely coherent since the beginning of the twentieth century, a period used to compare the variations, observed and predicted, of the average temperature of the planet. The results of an independent and new analysis conducted by a team led by the physicist Richard Muller, founder of Berkeley Earth Surface Temperature at University of California, Berkeley, were presented at the end of 2011. This team finds reliable evidence of a rise in world land temperature of approximately 1°C since the mid-1950s. As mentioned by Muller, this confirms that previous studies were done carefully and that potential biases identified by climate change skeptics did not seriously affect their conclusions.

Whereas the vast majority of scientists who are working on the recent evolution of our climate think that "warming is unequivocal," some climate skeptics have argued that in the last few years we have been in a period of cooling. Indeed, 2008 was colder than 2007—itself a bit colder than 2005—and warming has clearly shown a slowdown since 2001. But for a climatologist the climate must be assessed on the scale of decades at least. The conclusions are clear: the first decade of the twenty-first century was the warmest since the end of the nineteenth century, by a quarter of a degree compared to the 1990s, which were already considered warm. To look at the climate year by year is not without risk to the skeptics' arguments: 2010 was the hottest year in 130 years. The temperature data compiled by Jim Hansen's American team (NASA/GISS) support this assertion, even if there is a small difference between Hansen's data and those of the English team of Phil Jones

(HadCRUT). The difference observed between the two series, a tenth of a degree at most, can be attributed to the fact that Hansen used more stations in the Arctic regions. When the same stations are used, that difference is reduced to a few hundredths of a degree, an agreement that demonstrates, if need be, the futile nature of the attacks made against Jones within the framework of Climategate. Let's add that a slowdown such as that observed since 2001 is absolutely not abnormal in a climate that is warming under the effect of a regular increase in greenhouse effect. To be convinced of this, one need only examine in detail the projections made for different emission scenarios: warming is not regular from one year to the next, and the plateaus of a dozen years, or even more, can be identified in these simulations even if the greenhouse effect is regularly increasing. Finally, the idea, dear to skeptics, of a cooling under way makes no sense when we look at other indicators. Thus over the period 2005–8, Greenland and West Antarctica have each year contributed about a millimeter in the rise in sea level, and the sea level rise can only be explained by a warming effect (thermal expansion of the ocean, melting of continental ice). This is far from being anecdotal.

The difference between the warming measured on the surface of the Earth and that observed in the atmosphere from balloons and satellite data, which was claimed to be three times weaker, was another point on which skeptics based their arguments until recently. This difference was due to a skewing linked to the fact that the influence of the cooling of the stratosphere over satellite measurements was not correctly taken into account. Once that problem was fixed, the different sets of data provided results that were completely similar in the period over which they could be compared (1979 to the present), with warming by decade at 0.16 to 0.18°C on the surface and 0.12 to 0.19°C in altitude, leaving skeptics without an argument.

At least they couldn't argue about that point using recent data, but skeptics always appeal to the records from the last millennium. It is true that the warming of the last decades is on the same order of magnitude as the variations in temperature associated with the Little Ice Age or with the medieval optimum, both of which were of natural origin. However, the recent warming is global while those two events were more regional. To attribute the current warming to natural causes, in particular to variations in solar activity, as is often advanced by skeptics, does not hold up to scrutiny. During the last centuries, the variations in radiative forcing due to those of solar activity are close to ten times less than those due to human activity over the same period. It would thus be necessary for the effect of solar activity to be considerably amplified in relation to that due to our activities for the Sun to lead the way today. Let's imagine, however, that this has been the case; with the amplification working in the two directions, the year 2007 should have been relatively colder than five or six years earlier, since the eleven-year cycle that characterizes solar activity was at its minimum. This wasn't true. The year 2007 was the fifth warmest year, and this in spite of the cooling of part of the Pacific associated with La Niña. Another difficulty for the skeptics: a variation in solar activity should produce a rather homogeneous modification in temperature over the entire vertical thickness of the atmosphere, whereas with an increase in the greenhouse effect we expect a warming of the troposphere and a cooling of the stratosphere. And this is indeed what was observed; the cooling of the stratosphere is in a certain sense the signature of the role of human activity.

We hope we have been convincing: attributing the recent warming to human activity, described by the IPCC as very probable, is built on solid reasoning in which it is really difficult to find any holes. And when we turn to our white planet, the evidence that warming is occurring is abundant. Indeed, this was the key conclusion that was reached as an outcome of the debate organized by the French Académie des Sciences in September 2010 following the above-mentioned open letter signed by more than six hundred scientists involved in climate research. French academicians have concluded that "the greenhouse gas increase, a large part of which is due to human activities, plays an essential role in the current warming."

The White Planet on the Front Lines of Global Warming

When we examine the geographic distribution of the recent warming it appears that it has been less rapid at the surface of the oceans than on the continents, in fact twice as slow if we consider just the last two decades. The ocean has great inertia: it takes time for the additional heat to be transferred from the atmosphere to its surface layers. Thus it has trouble following the rhythm of warming, which explains a larger warming over the continents. This amplification is even greater in the regions of the Arctic, which during the twentieth century warmed on average twice as fast as the rest of the world, although in an uneven way.

One of the reasons for this polar amplification resides in the shrinking of the snow-covered surfaces of the Northern Hemisphere. At its highest level at the beginning of the spring, the snow cover, which reached 38 million km² in the 1920s, was henceforth closer to 35, with a significant dip in the 1980s. During the last twenty-five years these snow-covered surfaces have decreased by around 2 million km², about five times the surface of California; that amount of reflective surface was replaced by ground and vegetation, which absorb solar radiation. This change in albedo is far from being the only cause of polar amplification, as changes in the atmospheric circulation might also have contributed in a more important way.[14] The snow cover has also been modified in mountain regions, as is seen by a thinner layer of snow, by the melting of the earlier and less abundant snow, or by a rise in the line of equilibrium corresponding to the isotherm 0°C.

The ice cover of the Arctic Ocean has also decreased, as we mentioned at the beginning of this work, to a remarkable degree when we take into account the data from 2007. Let's recall that this ice is formed by the freezing of the seawater to a thickness of several meters at the most, generally less for ice formed during the year. The ice surface is smallest at the end of the summer, and that minimum level has decreased considerably. Between the beginning of the 1980s and 2005 it has gone from around 7.5 to 5.2 million km². And to everyone's surprise it suddenly decreased to 4.2 million km² in September 2007, a much more rapid decrease than any of the sea ice models had indicated (Figure 12.6). This unexpected acceleration of the disappearance of sea ice is probably the result not only of warming but also of changes in atmospheric circulation. It is accompanied by a very clear decrease in the thickness of that ice, of which a smaller proportion corresponds to ice formed over many years. Because young ice is more fragile, one might think that 2007 was a hard year for this ice cover, which would have trouble coming back in the coming years. However, this minimum level in the covering of ice increased in 2008 and then again in 2009. It then decreased, reaching in 2011 a quite similar small sea ice extent as that observed in 2007. And 2012 has been again a record year with a minimum extent of ~ 3.4 million km², considerably less than in 2007—a difference corresponding to about twice the size of California.

This change in the characteristics of the sea ice in the Arctic Ocean is also seen in the increased speed of the drift of the ice, as has been witnessed by the expedition of the sailboat *Tara*, which voluntarily became a prisoner of that

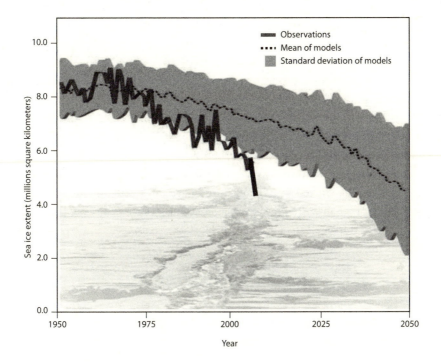

Figure 12.6. Arctic September sea ice extent: observations and model runs. This illustrates the minimum extension of the sea ice, which decreased very sharply in 2007, and the predictions of models. Source: National Snow and Ice Data Center, after J. Stroeve, M. M. Holland, W. Meier, T. Scambos, and M. Serreze, "Arctic Sea Ice Decline: Faster than Forecast," *Geophysical Research Letters* 34 (2007), L09501, doi:10.1029/2007GL029703.

ice in September 2006. Nine months later it had already traveled 1,000 kilometers, twice as rapidly as predicted, and came back to the open ocean at the beginning of 2008; its return had not been expected until the following summer. This acceleration in the drifting of the ice can moreover be held partly responsible for the decrease in the surface covered by the sea ice at the end of the summer of 2007. We noted earlier that this acceleration of the consequences of warming, particularly great in the Arctic, also seems to affect the coastal regions of Greenland. There, too, the recent data are extremely impressive because of the possible consequences in terms of sea level or other climatic events. The ice is flowing from the inside of the ice sheet toward the ocean through huge glaciers, true rivers of ice, at a flow rate that is already quite high but which, for some glaciers, has more than doubled during the last five years, reaching speeds of forty meters per day at the front of the glacier. That acceleration seems to be linked to two phenomena. The first is that the base of the glacier is lubricated and slides better as a result of the infiltration of liquid water linked to the melting that is accelerating in the coastal regions of Greenland. The second is the disappearance, here too due to warming, of certain ice shelves, which until recently were present at the mouth of these glaciers and slowed their flow into the ocean. Recent estimates, thanks to the interferometry radar work led by a team headed by Éric Rignot from the University of California,[15] have shown that in 2005 Greenland lost more than 200 billion tons of ice, around twice as much as it did in 1996. As we indicated in chapter 2, this loss of mass has continued and has even increased.

The acceleration of these glaciers is recorded by networks of seismometers placed to detect and record earthquakes. When these enormous flows accelerate then slow down, they generate low-frequency seismic waves that can be distinguished from those caused by a regular earthquake and enable us to follow at a distance the way in which the flow of Greenland is evolving. The results corroborate those obtained from the satellite observation of the glaciers. A team of American researchers, led by Göran Ekström,[16] has shown an increase in these mini glacial earthquakes since 2002. In 2005 we counted twice as many as in any other year prior to 2002.

Additional proof of this warming in high latitudes of the Northern Hemisphere is in the lakes and rivers that on average freeze later and thaw earlier than they did in the middle of the nineteenth century—by around six days in both directions. The permafrost is also affected, with a warming on the surface of more than 3°C compared to the temperature in 1980 and whose base thaws by a few centimeters every year. The same is true of the ground that freezes and thaws each year, whose surface has decreased by 7% since the middle of the twentieth century and whose maximum thickness has decreased by 30 centimeters.

And what about Antarctica? Unlike the Arctic, warming there is not as noticeable compared to the global average, probably because it is surrounded by a huge ocean and not by continents. There were no sufficiently viable data for the sea ice covering around the Antarctic continent before the satellite data were available at the beginning of the 1970s, and since then the minimum values of the sea ice covering, which are quite variable from one year to the next, do not show a significant change. In fact, the most visible events around Antarctica, whose coastal regions are colder than those of Greenland and thus a priori less sensitive to a warming of 1 or 2°C, concern the very thick ice shelves that carry the ice cap far into the ocean. Not the huge Ross and Filchner ice shelves, which are as large as France and not much in the news, but those that border the Antarctic Peninsula.

Among them is the Larsen B Ice Shelf, which was in the headlines in February and March 2002. Despite a thickness greater than 200 meters in some places, more than 3,000 km², close to the size of Rhode Island, has broken off and scattered in the form of enormous icebergs, which have then melted in the warmer regions of the ocean. Over the preceding five years that ice shelf had already lost almost the equivalent surface, and during the last few decades around 12,500 km² of ice shelves have disappeared. And this in a region where they were present since the beginning of the current Holocene period, around 10,000 years ago.[17] It is probable that this disappearance is linked to the particularly pronounced warming of the Antarctic Peninsula (by more than 2°C during the last fifty years), but it is tricky to make a connection between that occurrence and global warming. It is important to note that the gradual warming of the ocean waters circulating under such ice shelves also likely plays a role. Finally, the disappearance of the Larsen B Ice Shelf resulted in an acceleration of the glaciers in this part of the Antarctic Peninsula, whose flow toward the ocean is henceforth much easier. At the beginning of 2008 part of the Wilkins Ice Shelf, located at the southwest of the Antarctic Peninsula, broke apart.

Thanks to the work of Éric Rignot and his coauthors,[18] we are beginning to better assess the mass of the different basins of Antarctica over the recent period. Rignot's study, which covers the period 1992–2006, confirms that East Antarctica is in quasi-balance but shows very notable losses over West Antarctica and the Antarctic Peninsula, which have reached close to 200 billion tons in 2006. These figures are comparable to those that we have cited for Greenland. The phenomenon has amplified in the past few years such that Antarctica on the whole henceforth has a negative mass.

Wherever we look, with the exception of the sea ice surrounding the Antarctic continent and East Antarctica, all the components of the cryosphere— snow-covered surfaces, frozen ground, frozen rivers and lakes, sea ice, ice shelves, glaciers, and ice sheets—thus appear to be affected by climate warming, with a clear acceleration over the last decade. The consequence of this is a significant contribution of glaciers, ice caps, and ice sheets to a rise in the sea level; thus over the period 1993–2003 those factors are at the origin of nearly half of the three-centimeter observed increase in the water, the largest part of which is due to glaciers and ice caps.

The 400 billion tons that sum up the summer losses of the ice sheets, in 2005 for Greenland and in 2006 for Antarctica, have contributed to a rise in the sea level of more than one millimeter, and this over the course of just one year. Since then, the annual losses to the ice sheets have, on average, been greater.

CHAPTER 13

What Will the Climate Be in the Future?

Attributing climate change to human activity has been the subject of debates ever since the creation of the IPCC. It will take another few years, probably a decade, perhaps more, for this to become an uncontested assertion. Continuing to acquire quality data, better understand the role of aerosols, better identify the natural causes, know more precisely the sensitivity of the climate—in other words its reaction vis-à-vis a modification in the radiative forcing, as well as its natural variability and causes—are the directions in which we must collectively progress. However, there is a near certainty that we must face when we turn to the future: the climate will continue to warm. Why is this? Quite simply because even if, by waving a magic wand, we were able to instantly stabilize the greenhouse effect, the climatic system has an inertia such that it would still warm by an amount almost equivalent to what has occurred over the twentieth century. In fact, limiting warming to 2°C, even compared to the climate at the end of the twentieth century, is a true challenge, whose different facets we will examine in a later chapter.

Let's first look at a climatic future in which the need to stabilize the greenhouse effect would not be taken into account. This is in fact the process widely undertaken by the IPCC experts in the 2007 report of the scientific group, which we will look at broadly here for that which involves the global climate and in more detail for that which concerns our white planet: glaciers, sea ice, frozen ground. But we will also look at the sea level, whose rise is largely associated with the behavior of polar ice caps and inland ice sheets, and at the Gulf Stream, whose evolution is strongly influenced by that of the climate in high latitudes.

A True Upheaval if We Aren't Careful

The only way to predict the climate of the future is to use climatic models. Granted, there were periods in the past that were warmer than the current climate, for example, during the last interglacial period 125,000 years ago. But the origin of that warming related to a different amount of sun radiation has nothing to do with an increase in the greenhouse effect. That climate cannot therefore be used as an analogue of the future warmer climate, even if it is full of information, for example, about the evolution of the sea level.

These global climate models have been considerably improved since the work of the first climate modelers, who took only the atmosphere into consideration. The aim of such a model is to simulate the movements of the atmosphere but also to predict precipitation, evaporation, and more generally everything that has to do with the cycle of water. The basic equations for dynamics of the air, of physics for that which bears on the water cycle, are similar to those used to predict the weather. But whereas the meteorologist follows the evolution of perturbations from their origin until he loses any trace of them at the end of a few days, the climatologist looks at the average readings of temperatures, winds, and precipitation.

In the 1980s interactions between land surfaces and the atmosphere were integrated, whereas oceanic models were developed separately. At the beginning of the 1990s coupled models included the atmosphere, the continents, the oceans, and the ice, then they began to take into account the role of aerosols and the carbon cycle, and more recently the chemistry of the atmosphere and the dynamics of vegetation. There are currently more than twenty of these models throughout the world, including two in France, one at Météo France, the other at the Institut Pierre-Simon-Laplace; in the United States, General Circulation Models were first developed at Princeton (Geophysical Fluid Dynamics Laboratory [GFDL]), New York (NASA/GISS), and Boulder (National Center for Atmospheric Research [NCAR]). All these models differ; some emphasize the carbon cycle or the role of aerosols, while others focus on vegetation or on the interactions between the atmosphere and the ocean. This diversity is a rich source of information.

The confrontation with reality illustrates the quality of these climatic models developed during the last decades. Their ability to simulate broad characteristics of atmospheric circulation was quickly demonstrated. They take into account the rise of humid air in equatorial regions, the fall of dried air in subtropical regions where the great deserts are located, and the more horizontal circulation in our latitudes around depressions and anticyclones. We can verify that these models reproduce the climate remarkably well, not only during a year month after month, season after season, but that they are also able to describe the climate over longer periods. This is seen in the comparison of simulations of the climate of the twentieth century with variations observed over that period (Figure 12.4) or climates reigning over other planets: Mars and Venus, with such different conditions, or over our planet Earth during the extreme conditions of the Last Glacial Maximum. Or, closer in time, of those that prevailed six thousand years ago with a more pronounced cycle of monsoons in response to a sharp difference between the amount of sun in the summer and the winter.

Despite all of this progress, these climatic simulations, which require many computing hours, still convey some uncertainty. Thus our knowledge of climate sensitivity has scarcely progressed in thirty years. And yet there is a simple exercise, identical for each model, which consists of simulating the climate for a concentration of CO_2 multiplied by two. Between 1.5 and 4.5 in 1979, the range of the results has tightened only slightly; it goes from 2 to 4.5°C in 2007 with the most probable number of 3°C. Somewhat anecdotally, we were correct in indicating, as seen in the article published in 1990 conjointly with Jim Hansen and Hervé Le Treut, that the paleoclimatic data were consistent with a climatic sensitivity of 3 to 4°C. But more important is the fact that our scientific community henceforth has confidence in the lower value of that sensitivity, probably equal to 2°C and not lower than 1.5°C. There are thus indeed processes of amplification, connected first to water vapor, that are at work in the climatic system and not mechanisms of compensation such as those suggested by the American climatologist Richard Lindzen.

Lindzen defends, among other things, the idea known as "iris":[1] in tropical regions warming would provoke a decrease in the area of convective zones whose summit plays the role of greenhouse effect and thus an increase in those where the air dried by that convection drops, giving rise to a negative retroaction. Observations have not verified the validity of that hypothesis. It is, however, confirmed that the main source of uncertainty comes from retroactions linked to clouds. But we must also emphasize that this simple exercise of a doubling of CO_2 does not take into account all the sources of uncertainty, in particular those resulting from interactions between the carbon cycle and the warming that tends to decrease the intensity of absorption of carbon both on the continents and in the oceans.

To look into the future, the climatologist needs, in addition to models, to know the way in which the composition of the atmosphere and its greenhouse effect, as well as the other radiative forcings, those in particular connected to aerosols, will evolve. And here he turns to economists who have reflected on the evolution of CO_2, CH_4, N_2O, and other greenhouse gas emissions and that of aerosols from now to the end of the century. They define families of scenarios with different trajectories, obviously none of which aims to predict our future, rather the whole of which covers all possibilities. With a gamut, moreover extremely wide, taking into account the demography, the type of development, the way in which energy is produced and used, and so forth. Let's take the case of CO_2. The emitting scenario (Figure 13.1), A2, corresponds to a rapid increase based largely on the use of fossil fuel; in 2100, emissions would be multiplied by four compared to the beginning of the century, or 7 gtC, reaching nearly 30 gtC. The least emitter, B1, established on a society of services encouraging clean energy, is conveyed first by an increase followed by a decrease that brings CO_2 emission down to around 5 gtC in 2100. Not surprisingly, the most emitting scenario leads to the largest increase in concentrations that, in 2100, would be multiplied, compared to their preindustrial value of 280 ppmv, by around three; in terms of radiative forcing, taking into account the other greenhouse gases and aerosols, we would be at nearly 9 Wm^{-2} compared to 1750, an additional radiative forcing of around 4%.

Although it shows the least amount of carbon, scenario B1 leads to concentrations close to 550 ppmv, an increase of nearly 90%. Furthermore, these calculations do not take into account the interactions between the carbon cycle and the climate, which are likely to increase those concentrations from around 100 ppmv to 300 ppmv in an extreme case. More disturbing is the fact that this scenario does not lead to a stabilization of the greenhouse effect by 2100. Indeed, it is not sufficient to maintain emissions at their current level—which on average is more or less the case in this scenario—to reach the goal of stabilization of concentrations. This is implacable logic, since each year vegetation and oceans absorb only around half of what is emitted, in-

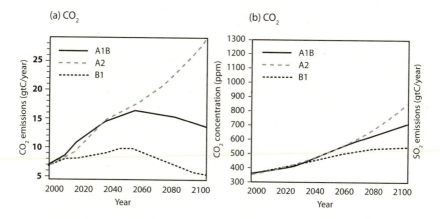

Figure 13.1. Scenarios from the IPCC: emissions (left) and carbon dioxide concentrations (right). Source: IPCC report, 1996.

deed a bit less; the other half accumulates in the atmosphere. In fact economists do not seriously envision that our emissions can diminish rapidly or greatly, a goal that is indispensable from a climatological point of view, which we will see below.

However, this range of scenarios enables modelers to make climatic projections that cover all of the climatic possibilities if we do not stabilize the greenhouse effect—unless something unexpected occurs (we will return to this). Let's look first at the average warming that will be reached by the end of the twenty-first century: the projections presented in 2007, which, for the first time, took into account the interactions between the climate and the carbon cycle, cover a wide gamut between 1.1 and 6.4°C, little different from those of the 2001 report. About half of the large range of these projections comes from the behavior of our societies, which must expect greater warming as they will emit more greenhouse gases, and the other half comes from uncertainty linked to the behavior of the climatic system. However, in 2007 we had more confidence than we did in 2001 in the central levels of the different scenarios, from 1.8°C for the most conserving to 4°C for the most emitting.

Let's take the case of an intermediary scenario, such as A1B characterized by emissions of CO_2 that have returned to 13 gtC in 2100 after having culminated at 17 gtC in 2050; the climate of the period 2080–99 will be warmer by close to 2.8°C compared to 1980–99 and by 3.3°C compared to the pre-industrial climate (Figure 13.2). We can easily see that this would be a true climatic upheaval, since the average warming that our planet would experience from now to the end of the century would be around half of that which characterized the passing from the last glacial period to current conditions, a path that has taken more than five thousand years.

This is especially true since the warming, which as we have emphasized has become more rapid, would be accompanied by other important climatic changes. Thus it is probable that heat waves and events of heavy precipitation will continue to become more frequent; the summer of 2003, which in France was more than 3°C warmer than the average summer of the twentieth century, would be considered a normal summer in the second part of the twenty-first century because France, like other continental regions in general, will warm more than the average. It is probable that tropical cyclones will become more intense with stronger winds and precipitations. Outside the tropics, it is predicted that the trajectories of storms will be more poleward, inducing changes in the distribution of winds, precipitation, and temperatures. The amount of precipitation will increase in the high latitudes and will decrease in the subtropical regions and as far as the Mediterranean basin, with less rain in the summer over a large part of France, up to 30% less rain in the south.

Beyond this very general overview, everything that involves the snow and ice will be greatly affected.

What Will Become of Our Glaciers?

Given the general retreat of glaciers during the second half of the twentieth century, with the exception of those of Scandinavia and New Zealand, it is legitimate to raise questions about the survival of some of them in a climate that would warm up notably during the twenty-first century. From the 2007 IPCC report we learn that the annual loss through ablation—essentially summer melting—is great, from 30 to 40 centimeters per degree Celsius, and that it could not be compensated for by a potential increase in precipitation that on average does not exceed 1 to 2%. With a few rare exceptions, the glaciers will thus continue to retreat, especially since the pollution lightly blackens the surface and from that decreases the albedo, as much as 10% for

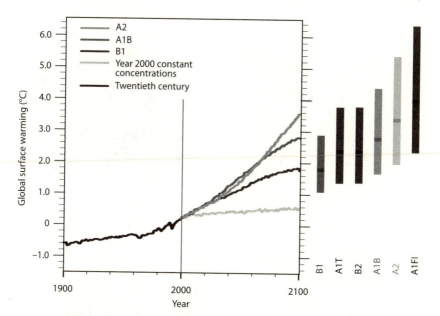

Figure 13.2. Evolution of the average global temperature for the twentieth century and projections for scenarios A2, A1B, and B1. The rectangles on the right indicate the warming predicted for a group of six scenarios ordered by increasing levels of emissions and the uncertainties associated with them. Source: S. Solomon et al., *Contribution of Working Group I to the Fourth Assessment Report of the Intergovernmental Panel on Climate Change* (Cambridge: Cambridge University Press, 2007).

some very polluted regions of the Northern Hemisphere. Beyond these general considerations, it is clear that the future of a glacier depends on the regional characteristics of the evolution of the climate and of those of each glacier—its surface area, volume, orientation, altitude, and geometry of flow. They must be addressed on a case-by-case basis.

Once again we will rely on the extremely well-documented work by Bernard Francou and Christian Vincent, who undertook this inquiry, and take a few examples from it. They draw attention to the fact that for a given warming the annual melting depends enormously on the altitude: in the French Alps, warming has a determinant effect in the lower part of the glacier (up to 2,400 meters) and less or not at all in its upper part (around 3,600 meters and above). The glaciers that are most sensitive to the climate are the

Temperature

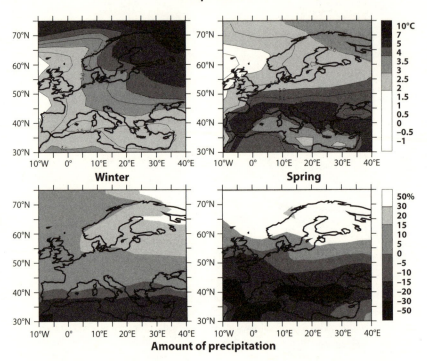

Amount of precipitation

Figure 13.3. The top two figures correspond to the warming predicted for Europe between the periods 2080–99 and 1980–99 in the case of the median scenario A1B, in winter (left) and in summer (right). The bottom two figures indicate variations in the amount of corresponding precipitations. Source: S. Solomon et al., *Contribution of Working Group I to the Fourth Assessment Report of the Intergovernmental Panel on Climate Change* (Cambridge: Cambridge University Press, 2007).

maritime glaciers of Scandinavia and New Zealand. With moderate warming, the altitude of the line of equilibrium should rise to around 300 meters in the Alps, which would condemn glaciers like Saint-Sorlin, except for a bit of residual ice not at the summit but in a valley located at 2,900 meters, of Gébroulaz, and that of Arolla in the Valais. That warming would leave little chance of survival for the Argentière Glacier; those of Nigardsreen in Norway and Franz-Josef in New Zealand, which are distinguished by strong advances at the end of the twentieth century, in a hundred years will be severely reduced in the case of the former and a bit less so in the case of the latter. The Mer de Glace should resist better due to the high altitude of its basin of accumulation. If we add shorter snow seasons to that, mountain massifs around the world will be quite different from what they are today. And this is also true for those we have not explicitly cited in North America, South America, Africa, and Asia.

An Arctic Ocean without Ice?

The contrast is astounding when we compare the warming projected in the Arctic on the one hand and in Antarctica on the other. Of the same magnitude as the average warming in the south—around 3°C in the A1B scenario—it is amplified by a factor of two, even three, in the high latitudes of the Northern Hemisphere where regionally it could reach 10°C. These high values are in line with those that have been observed in the Great North for a few decades and result from the processes of amplification that we have already discussed. They are accompanied logically by a decrease in snow cover and a melting of frozen ground with an increase of 30 to 40% of thickness in the active zone, which thaws then freezes every year, and by substantial increases of methane emissions in those regions.

One of the great questions raised by warming concerns the sea ice and its disappearance in the summer; that disappearance would enable some maritime routes to open—initially just partially then eventually extending to the entire Arctic Ocean. Part of the year it would then be free of ice. Here, too, the positive retroactions come into play. Thus when sea ice disappears, the ocean surfaces that replace it are more absorbent, which tends to increase melting. The 2007 IPCC report predicted a moderate decrease in this winter ice and an almost complete disappearance of it in the summer in 2100. This decrease would be less great around Antarctica than in the Arctic. The report also noted a disparity in the results obtained by different sea ice models. The results of the American model from the NCAR, which projects a disappearance of the summer ice by 2040, were emphasized as soon as the summer observations of 2007 that demonstrated an extremely abrupt decrease in the extent of ice in the Arctic were published. In retrospect (Figure 12.6), none of the models used had foreseen such a decrease, and some researchers, followed by Al Gore, went so far as to predict the definitive disappearance of the ice during the summer in the next ten years or even sooner. The rise in the minimum ice cover in 2008 and then in 2009 calls for caution in this type of prediction, but the very low value observed in 2012 gives some credit to it.

Surprises under the Frozen Ground

When you walk along the prestigious Nevsky Avenue in Saint Petersburg, you discover in the souvenir shops many objects sculpted out of mammoth ivory from the remains of those mammals that disappeared ten thousand years ago and were rediscovered in the frozen ground of Siberia. Riches of the frozen ground, mammoths were not the only things found. In particular, that ground contains methane hydrates that are particularly sought after and desired by oil and mineral companies because they are a potential source of fuel that can replace other fossil fuels (oil, coal).

What are these hydrates (also called clathrates) of methane? Are they as mysterious as suggested by the title of the book by Gérard Lambert and his colleagues, *Le méthane et le destin de la Terre. Les hydrates de méthane: Rêve ou cauchemar*? In this book the authors describe the state of the art of these strange substances. As their name indicates, their formation necessitates at least two basic ingredients: H_2O molecules and methane molecules, CH_4. But the intimate blending of these molecules requires very special conditions because H_2O molecules must be organized in a sort of cage within which those of methane are imprisoned, hence the name clathrate, from the Greek *klathron*, which means closure. Two factors are necessary for the sequestration of the methane and for the endurance of its hydrates: high pressure and low temperatures, which explains their presence in the frozen ground.

For that which involves the constituents of the cage, the ground generally contains water, which freezes when the temperature becomes negative. In the polar regions this very deep ground remains permanently frozen—we call this permafrost, whose thickness can reach more than a kilometer in Siberia. Near the surface—up to several dozen meters depending on the site—the active layer melts in the summer and freezes again in the winter. The permafrost, not covered with glaciers or ice caps, is essentially concentrated in the Arctic where it covers a huge surface.

This ground, as in the case of frozen peat, is rich in organic matter. Methane, the principal component of natural gas, is formed by methanogenic bacteria that live in an anaerobic environment (that is, deprived of oxygen) from the process of decaying and of digestion of organic matter, whether of vegetal or animal origin. Thus in the permafrost, the source of methane can come from the presence in situ of methanogenic bacteria. But the analysis of this methane indicates that for the most part it comes from the migration of sources as deep as five thousand meters where it is formed at high temperature—we then speak of thermal methane—when buried vegetal remains begin to carbonize under the effect of heat to form peat, lignite, and coal and to free methane. Are the hydrates of methane in the permafrost a dream or a threat for the future?

They are a dream because their use is a potential source of energy whose combustion emits less CO_2 and fewer atmospheric pollutants than that of oil or coal. Estimates show a huge reservoir of methane hydrates on the Earth, but the difficulty and costs of extracting it are on par with the amount of the reserve, and most hydrates are found at the bottom of oceans; frozen ground contains only a small proportion of them.

But they are also a threat for the future insofar as climate warming is concerned. As indicated in the 2007 IPCC report, the area occupied by the permafrost in the Northern Hemisphere could decrease by 20 to 35% by 2050, essentially in the more southern zones. Furthermore, the thickness of the active layer, the one that melts each summer and freezes the following winter, will increase in many places. These two phenomena will contribute to a surplus of methane emissions into the atmosphere. However, the IPCC report remains extremely cautious compared to the hypothesis of an increase in the sources of carbon dioxide in Arctic regions.

A More Rapid and Higher Sea-Level Rise than Predicted

We have deliberately left aside aspects connected to a future rise in sea level. Granted, only part of the rising is connected to glaciers, ice caps and inland ice sheets; the other results from thermal expansion. Both are important in the evolution of the sea level in the mid- and long term, but the largest uncertainty lies in the contribution of ice.

What are the facts regarding the dilatation of the oceans? It is directly connected to how much warming occurs. From the least- to the most-emitting scenario, the average value of thermal expansion in the twenty-first century increases from 17 to 29 centimeters with, for a given scenario, a range of variation of more than a factor of two.

The contribution of the cryosphere is analyzed separately for the glaciers and the polar ice caps, which have contributed between 10 and 12 centimeters to the rise in sea level. As for Greenland and Antarctica, the difficulty lies in correctly evaluating the difference in the surface mass and the flow of ice that is emitted into the ocean, which depends on the dynamics of the glacier. Models as well as data indicate that the warming of the high latitudes of the Northern Hemisphere is accelerating the melting on Greenland in the coastal regions. The increase in snowfall that results from an intensification of the hydrological cycle is not able to compensate for this increased melting (the warmer it is, the more likely a cloud is to contain water vapor and to provide precipitation). The IPCC estimates that over the entire twentieth century, Greenland contributed an average 3- to 6-centimeter rise in sea level. Antarctica, on the other hand, would be too cold to be affected in a significant way by melting in coastal regions; on the contrary, it could increase in volume thanks to that increase of snowfall and thus very partially stall the rise in sea level by 6 to 8 centimeters on average. The result of all of these contributions is an average rise in sea level of 28 to 43 centimeters, which becomes even greater since we know that warming will increase, which means that, taking into account the different uncertainties of the parameters, the rise could be approximately 18 to 59 centimeters.

But these estimates do not fully take into account the acceleration observed in the flow in the coastal regions in Greenland, as well as in West Antarctica. This acceleration has not been formally attributed to human activity. But the IPCC's preliminary estimates show that if it continues alongside warming it could cause an additional 10- to 20-centimeter sea level rise from now to the end of the century. Some researchers argue that the rise in sea level is underestimated and that it could exceed one meter by 2100.[2] What has been observed during the last few years should caution us not to reject those alarmist estimates without another form of analysis; following the rhythm observed for 2005 and 2006, Antarctica and Greenland would contribute to a rise of about 10 centimeters by 2100, not taking into account the IPCC estimates. This contribution could be even greater if the flow of the emitting glaciers continues to accelerate with climate warming.

The IPCC's caution is probably justified; thus the data on a glacier in

Greenland show that after a rapid acceleration the glacier slowed considerably the following year whereas in West Antarctica the flow of the Rutford Glacier affected by the rhythm of the tides can vary by 20% over short periods (around fifteen days), which shows how difficult it is to deduce the evolution in the long term from measurements that are generally fixed at a given moment in time.[3]

There are uncertainties, then, about sea-level rise from now to the end of the century, which will be greater than the 17 centimeters observed during the twentieth century; the rise could be as much as sixfold if we believe the most alarmist projections. But what was emphasized in the IPCC's synthesis report was the irreversible nature of that rise beyond the twenty-first century owing to the inertia of the climatic system (Figure 13.4) and of the threat of a partial melting of the Greenland ice cap and possibly of the ice cap of West Antarctica.

The inertia of the thermal expansion is much greater than a few decades, which is characteristic of the equilibrium of temperatures between the atmosphere and the upper ocean. This last thermal equilibrium concerns only the surface layers of the ocean, whereas dilatation is produced over the entire water column. It takes a lot of time for the heat to be transferred to the ocean depths; one then measures in centuries, even in millennia. Consequently, even if radiative forcing and the temperature are stabilized, the rise in sea level will continue: from 30 to 80 centimeters in 2300 in the hypothesis of the median A1B scenario, followed by a stabilization of the radiative forcing in 2100, then at a lesser rhythm over many centuries.

The current models, which are still very preliminary, indicate that Greenland will ultimately be eliminated, which would cause a rise in sea level of around seven meters if the average temperature of the atmosphere were, for millennia, maintained above a certain level. That level is not in fact very high since the estimates vary from 1.9 to 4.6°C compared to the preindustrial climate. Another reason to be worried lies in the fact that these temperatures are comparable to those of the last interglacial period, 125,000 years ago, during which the volume of ice in the polar regions was less and the sea level four to six meters higher. Antarctica will remain too cold to melt significantly. However, its mass could decrease, thereby contributing to the sea level, if the flow connected to the dynamics of the ice accelerates; this would affect West Antarctica in particular.

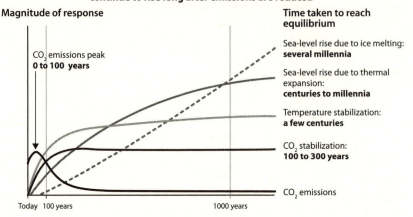

Figure 13.4. The inertia of the climatic system. The concentration of carbon dioxide, the temperature, and the sea levels continue to increase after emissions have reached maximum levels. Source: IPCC report, 1995.

Over a few centuries the rise in sea level could thus reach three to four meters, assuming that the volume of Greenland diminishes by a third. Even if this time frame is distant, it should cause us to reflect because great zones would be forever under the sea, and the wealth and beauty of many coastal cities would be lost forever. The situation is all the more urgent because it is our activity in the coming decades that will determine the warming at the end of the century and the sea-level rise in the centuries to come (Figure 13.4).

The Halt of the Gulf Stream

There are two reasons to look at the questions raised by the possible halt of the Gulf Stream. The data obtained by glacial ice cores in Greenland have largely been at the origin of the discovery of rapid climatic variations that we believe are connected to the halt then the starting up of the Gulf Stream, with abrupt changes in temperature that could have reached 16°C at the center of the ice sheet in a few decades. Based on these data, the 1995 IPCC introduced the notion of "climatic surprise." The processes that threaten to lead to a halting of the Gulf Stream in the future differ from those that operated in the past. But the simple fact that rapid and large variations have henceforth been well documented give a certain credibility to the fears that some people—climatologists, writers, filmmakers, politicians, and so forth—have for the future of the Gulf Stream, to which we attribute, even if it is only partially true, the mildness of the climate in regions from Western Europe under the influence of the Atlantic Ocean.

There is a common characteristic, nevertheless, between the past and present and future operation of the Gulf Stream. Just as it has been in the past, the Gulf Stream is sensitive to the arrival of freshwater into the surface waters of the North Atlantic. That current constitutes the northern extremity of what oceanographers call inverse thermohaline circulation, which goes along the western coasts of the North Atlantic; the term *thermohaline* indicates that this circulation is affected by both the temperature and the salinity of the surface ocean waters. In almost all climatic models, warming is accompanied by a notable increase in temperature and precipitation in the high latitudes of the Northern Hemisphere. Each of these effects—warming and an additional amount of freshwater (either directly for ocean precipitations or through an increase observed in the contribution of rivers flowing into the sea in these regions)—tends to make the surface waters of high latitudes less dense. They then have increasing difficulty sinking into the deep water.

None of the simulations that couple general circulation models of the atmosphere and of the ocean (or even less complex models) points to an abrupt stop of the thermohaline circulation during this century. They suggest a slowing instead. But it is important to note that the regions of Europe under the influence of the North Atlantic continue to warm. However, some models, generally those of "intermediary complexity," suggest that the thermohaline circulation and thus the Gulf Stream could stop in the longer term in response to sufficiently increased warming.

Simulations using coupled complex ocean-atmosphere models also suggest that in the next five hundred years the thermohaline circulation might indeed completely stop,[4] once the contribution of freshwater connected to the melting of Greenland is taken into account, which in the long term threatens to be sufficiently great to have an impact. The same simulation carried out without taking that contribution into account indeed does not lead to a stop in thermohaline circulation.

Above all we must not imagine that this halt will in any way provoke a planetary cooling. The average warming is scarcely less marked in a simulation leading in the long term to a halt in the thermohaline circulation than in the experiment in which thermohaline circulation remains active, 3.2 compared to 3.5°C, both of which correspond to a doubling of the concentration of carbon dioxide. And in the case of a halt of the thermohaline circulation, the warming would be less in the North Atlantic and in the Arctic, but it would be little affected in Western Europe. Catastrophic scenarios, as in the film *The Day after Tomorrow*, predicting the beginning of a glacial period resulting from a weakening of the thermohaline circulation are thus mere speculations.

Let us end this chapter on a more anecdotal note, although it involves a Pentagon report. In 2003, upon the request of the American military, Peter Schwartz and Doug Randall,[5] two consultants unknown to the community of climatologists, imagined that an abrupt event similar to that of 8,200 years ago could rapidly occur. The halt in the thermohaline circulation, described in their chapter devoted to the climate of the past 10,000 years, was, in Greenland and in the regions near the North Atlantic, first conveyed by a notable cooling—as much as 5°C—in a few decades. The main idea of this report was that we would be very close to the threshold from which the thermohaline circulation, which is indeed slowing down, threatens to stop—in 2010. With serious droughts in southern China and northern Europe during the following ten years, and, after a decade of cooling, the European climate would be close to that of Siberia. Famine would then spread in a cold and starving China that would jealously covet the energy resources of Russia, and chaos would be unleashed there with civil and border wars between 2020 and 2030. These geopolitical aspects and their consequences for the national security of the United States were at the heart of the report: climate refugees, the risks of war in the Caribbean and in Asia, internal struggles among European countries, new countries developing nuclear weapons, and so forth—a world of conflicts that would justify strong diplomatic action on the part of the United States.

There is nothing really serious in this study, which is without any scientific credibility and whose vision is too apocalyptic. Moreover, at least one of its predictions did not occur: the authors imagined that in 2007 the waters could breach the dams of the Netherlands, rendering The Hague uninhabitable. It is nevertheless true that even in the warmer climate that will prevail in the middle of the next century, a rapid cooling associated with a halt of the Gulf Stream would be extremely destabilizing from an ecological, economic, and geopolitical point of view, even if no country saw its temperatures go below those that it currently is experiencing.

But this is not the problem our societies will have to face between now and the end of the century or probably beyond that; it is a climate warming toward which we are inexorably headed.

CHAPTER 14

A Warming with Multiple Consequences

It is rather common, sometimes rightfully so, for our community to be criticized for being catastrophists when we broach aspects connected to the consequences of climate change. Alongside our knowledge of the climatic system and its evolution, that of the impacts that a warming would have has progressed greatly in the last twenty years. We now also rely largely on the 2007 IPCC Group II report, which is dedicated to the impacts of climate change and to our adaptation and vulnerability to it.[1] Thanks to that synthesis, we will reveal the world toward which we risk evolving if nothing is done to limit greenhouse gas emissions. Before turning to the glaciers and polar regions, we will take a quick look at consequences on a global scale.

A True Upheaval on a Global Scale

Intuitively we expect that the consequences of climate change will increase as the temperature rises; they will also depend on the evolution of climatic extremes that contribute in an important way to the damage. This is indeed the case, as is seen at point 1 in table 14.1, which presents a synthesis of the main consequences of climate warming according to large sector: water, ecosystems, food, coastal areas, and health (Table 14.1). Thus the threshold of 1.5 to 2.5°C appears critical for the maintenance of biodiversity—greater warming leads to important changes in the structure and functioning of ecosystems—and in the ability of continental ecosystems to play their role of carbon pumps. In the ranks of particularly vulnerable ecosystems are the tundra, the boreal forest, ecosystems of mountain regions sensitive to warming, and those in the Mediterranean or certain forests in tropical regions as a result of the decrease in precipitation.

Table 14.1

Predictions regarding the impacts of climate change in various realms
in function of the amount of warming.

Global average annual temperature change relative to 1980-1999 (°C)

	0	1	2	3	4	5
WATER		Increased water availablility in moist tropics and high latitudes - - - - - - - - - - - - - - - - - - - ▶				
		Decreasing water availability and increasing drought in mid-latitudes and semi-arid low latitudes - - - - - - - - ▶				
		Hundreds of millions of people exposed to increased water stress - - - - - - - - - - - - - - - - - - - ▶				
ECOSYSTEMS			Up to 30% of species at increasing risk of extinction ——— Significant extinctions around the globe ▶			
		Increased coral bleaching —— Most corals bleached ——		Widespread coral mortality - - - - - - - - - - - ▶		
		Increasing species range shifts and wildfire risk	Terrestrial biosphere tends toward a net carbon source as: ~15% ———— ~40% of ecosystems affected ▶			
			Ecosystem changes due to weakening of the meridional overturning circulation - - - - - - - - - - - - ▶			
FOOD		Complex, localised negative impacts on small holders, subsistence farmers, and fishers - - - - - - - - - - - - - - - ▶				
		Tendencies for cereal productivity to decrease in low latitudes ————		Productivity of all cereals decreases in low latitudes - - - - ▶		
		Tendencies for some cereal productivity to increase at mid- to high latitudes ————		Cereal productivity to decrease in some regions - - - - - - ▶		
COASTS		Increased damage from floods and storms - ▶				
				About 30% of global coastal wetlands lost[†] - - - - - - ▶		
			Millions more people could experience coastal flooding each year - - - - - - - ▶			
HEALTH		Increasing burden from malnutrition, diarrheal, cardiorespiratory, and infectious diseases - - - - - - - - ▶				
		Increased morbidity and mortality from heat waves, floods and droughts - - - - - - - - - - - - - - - - ▶				
		Changed distribution of some disease vectors - ▶				
				Substantial burden on health services - - - - - - - ▶		
	0	1	2	3	4	5

†Based on average rate of sea level rise of 4.2 mm/year from 2000-2080

Source: IPCC, *Climate Change 2007: Fourth Assessment Report* (Cambridge: Cambridge University Press, 2007).

Initially warming would not pose any problem vis-à-vis the potential for food production, which should increase if that warming remains locally inferior to a threshold of 1–3°C. But beyond 3°C agricultural productivity would decrease. In the low latitudes agriculture would also be affected because of a decrease in the availability of water resources, and agricultural production could be severely compromised in many African countries, reducing food security there even more and exacerbating malnutrition.

The problem of water resources, which even without a modification in our climate is already critical in many regions, risks being seriously exacerbated by warming. Indeed, rain would be less abundant in already dry regions of the mid-latitudes and the tropics. Combined with greater evaporation, this would lead to a decrease of runoff by 10 to 30%. The decrease in water resources due to climate change would also affect the semi-arid zones in the Mediterranean basin, the western United States, South Africa, and northeast Brazil. The regions suffering from droughts would thus extend with impacts on many sectors; the IPCC report anticipates that in Africa in 2020, 75 to 220 million people will be exposed to a hydro-stress due to climate change, and in 2080 the surface of dry and semi-arid land will have increased from 5 to 8%. Inversely, it will rain more where there is already sufficient rain (10 to 40% more in the high latitudes and in certain humid tropical regions), with additional risks of flooding tied also to more intense precipitation, including in regions where the total amount of precipitation will diminish. It is possible that in 2080 as much as 20% of the world's population will live in regions where the risk of flooding will be greater than it is today.

Coastal zones are particularly at risk because of the average rise in sea level but also because of an increase in the severity and frequency of storms, resulting in an increase in coastal flooding, especially since in some cases there is an added increase in overflow from rivers that flow into the sea in those regions. One thinks first of all of the consequences on populations of small islands that are particularly vulnerable and on the much larger populations of the megadeltas in Asia from India to Southeast Asia. A sea-level rise of 40 centimeters would cause the number of people whose homes are flooded each year, in particular in India, Bangladesh, Vietnam, and China, to increase from 13 to 94 million. The same is true in Africa, especially the very populated deltas of the Nile and the Niger, but other regions such as Europe and Australia will be affected. The third IPCC report[2] indicates that a rise of 40 centimeters would affect around 200 million more people, who would be forced to leave their communities; that number would be half if protections are put into place. In addition, in these regions there would be problems related to the erosion of the coasts, the salinity of water, and the fragility of certain ecosystems, mangroves, and humid zones.

When we mention "climate refugees,"[3] it is logical to turn to the regions that are potentially sensitive to a rise of a few dozen centimeters of elevation in sea level: Bangladesh, the Tuvalu Islands in the Pacific, the Maldives in the Indian Ocean, and the Halligen in the North Sea. But the threat of exceeding 40 centimeters between now and the end of the century is, as we have seen, far from negligible. And within a few centuries, we will be measuring in meters the cumulative effect, very slow but inevitable, of the dilatation of the ocean and the much more uncertain but highly threatening rise that might result from a partial melting of Greenland and even West Antarctica. When Susan Solomon, co-president of the scientific group, presented the results of the fourth report of the IPCC, she added a map showing what would happen to Florida if the sea rose six meters—or rather what would remain of Florida: not more than half, as all the coastal zones would be underwater. Al Gore used this same map in *An Inconvenient Truth*. Let us repeat: this could only happen within a few centuries, but the more greenhouse gas we emit in the next decades, the greater the risk that a situation of this type might occur. Four or five hundred years is far off, but it is on that scale of time that all these coastal cities with their inestimable architectural wealth that would then be impossible to protect were constructed. This long-term dimension of climate change should also give us pause.

Another problem that is mentioned more and more vis-à-vis the ocean is that of the coral in warm seas; it is estimated that 30% or more has disappeared in the last twenty years due primarily to an increase in the sea surface temperature. Most corals could die of bleaching as soon as the average temperature becomes warmer by more than 1.5°C, with consequences that are not just aesthetic: coral is indispensable to the survival of very rich ecosystems, including countless other species, many of which have not even been inventoried yet, and consequently, in many regions of the planet, to the survival of local activities with socioeconomic implications, notably tourism and fishing. An increase in temperature is not the only threat to corals. We must also add the risk of an increase in intensity of tropical cyclones and the acidification of the oceans, which also threatens the coral reefs in the cold seas. This acidification, linked to the ocean's absorption of a portion of the carbon dioxide emitted by human activity, has already been observed and should increase during the twenty-first century. Independently of the climatic role of CO_2, this acidification is a problem, the consequences of which for the marine biosphere have been little studied.

Warming will also have consequences for human health, including the risks mentioned already of malnutrition and those linked to the geographic modification of zones of influence of some vectors of infectious diseases or to a rise in cardiovascular illnesses resulting from an increase of ozone in the troposphere (itself influenced by climate change). Finally, events such as heat waves but also flooding, droughts, cyclones, and fires can cause death, illness, or injury. Of course, less rigorous winters in our regions can have positive consequences on the health of populations, but globally the negative aspects will dominate.

No region will be spared from the adverse consequences of climate change, but some are a priori more vulnerable. We have already cited them: areas that are particularly sensitive to the sea level rise, small islands, and megadeltas in Asia and Africa. And the African continent, with its weak capacity for adaptation, is at risk of being affected by other impacts. Another vulnerable region, the Arctic, is also at great risk because of the ever-increasing speed of climate warming. But before looking at the polar regions, let's turn to the mountain regions, which are also at the forefront of those at risk of being affected by climate warming.

Mountain Regions

We are often asked this question: What will the consequences of warming be in France? We begin by pointing out that by around 2050 the summer we had in 2003 will be an average summer; then we mention the aspects connected to the coastal regions, some of which will be affected by the elevation of the sea level, to agriculture, to forests, to the economy, to health, and to tourism. All these aspects are discussed in a series of articles published by Greenpeace with a view toward establishing a synthesis on the impacts of climate change in France.[4] But two consequences are usually pointed out. The first is connected to water resources: as is the case with the entire Mediterranean basin, they will be affected in the southern part of France. There will be less precipitation there, too, especially in the summer, and more evaporation. The second concerns our mountains. The impact of climate change will be very visible there, as illustrated by two articles published in the Greenpeace report. If warming is greater than 3°C, most of the glaciers in the Alps and the Pyrenees will be reduced to nothing. Only the largest glaciers located above 4,000 meters in the Mont-Blanc Massif will be spared but with a marked reduction of their areas and lengths. Even a 2°C warming would have notable effects on the conditions of snow cover at mid-altitude (1,500 to 2,500 meters). Thus at 1,500 meters, the number of months of snow cover would be reduced from five to four in the Northern Alps and from three to two in the Southern Alps and the Pyrenees. This would equate to a decrease of 40 to 50% in the snow cover. In high altitudes the snow cover would be reduced by about ten days, but that decrease would be more pronounced in the event of greater warming.

Beyond the consequences for winter tourism—the very existence of winter resorts and recreation areas would be threatened—the decrease in snow cover would affect other areas as well. Spring avalanches could become more frequent, and the flow of the rivers would likely change, with the peak of the spring melting occurring about one month earlier, which would affect irrigation and hydroelectric resources. We should expect a gradual rise in the upper limit of the tree line, thus limiting the zone of alpine plants. In addition, the disappearance of frozen ground associated with warming can result in sliding ground on mountainsides and falling rocks. What is true for the French mountains can be more or less applied to the entire Alpine massif and to other mountainous regions and thus affect tourism and recreational activities. For example, without snowmaking, the ski season in western North America would likely shorten substantially, with projected losses of three to six weeks (by the 2050s) and seven to fifteen weeks (2080s) in the Sierra Nevada of California (for high emissions IPCC scenarios).[5] The African and some glaciers in the Andes will also be subjected to a harsh trial, in particular the smaller ones like Chacaltaya in Bolivia, which has recently disappeared. The situation is particularly alarming in South America, where many artificial lakes used for irrigation are filled almost exclusively by glacial melting.

Polar Regions: Multiple and Diverse Impacts

Given the expected differences when faced with an increase in the greenhouse effect—the Arctic's warming is two to three times greater than the global average while that of Antarctica is much closer to the global average—it is not surprising that the impacts would be very different for the northern and southern polar regions, especially since each region varies in terms of fauna, flora, and populations.

The ecosystems are already sensitive to the changes that are occurring. For example, the abundance of krill, small shrimp that play an essential role in the food chain and are a key element in ecosystems in Antarctica, is decreasing in the southern waters, as are colonies of Weddell seals and penguins. The populations of krill and penguins vary in strict correlation with the sea ice because penguins feed mainly on krill, whose reproduction depends on the expanse of sea ice. In the Weddell Sea and in the Antarctic Peninsula, warming, which has been steadily increasing over the past fifty years, has meant a reduction in the area of the sea ice in winter and spring and thus a decrease in the number of krill and Adélie penguins by 30% in twenty years. The penguins cannot reproduce except when the sea ice surface is at a maximum, which allows us to predict their disappearance on the Antarctic Peninsula by the end of the century. Another study conducted on the Crozet archipelago indicates that a warming of only a few tenths of a degree constitutes a serious threat for the king penguin.[6]

Other works show that the extent of the ice also determines how marine ecosystems function in the Arctic, especially since they do not have the opportunity to adapt to the very rapid changes that are occurring. On the emerging land, some species—the Arctic fox, wild geese—have decreased greatly while others—élans, some birds—have moved to the north. Vegetation has already been modified in some regions and projections indicate that in the case of moderate warming 10% of the tundra will be replaced by forest.

A decrease in the surface area of the sea ice and its thinning will have significant consequences for certain species. Polar bears, for example, need the ice shelf as a platform to hunt seals, their main food source. The population of polar bears, which can travel thousands of kilometers per year in the regions around the Arctic, is estimated at 20,000 to 25,000, but the survival of the largest living land carnivore is in danger because of the shrinking of its habitat; polar bears, moreover, have been placed on the red list of endangered species by the International Union for the Conservation of Nature (IUCN). The first signs of decline have been observed in the Hudson Bay in Canada; a decrease of more than 30% in the next thirty-five to fifty years is anticipated, and there is a serious risk of extinction if the sea ice shrinks greatly.

Four million people, 10% of whom are native, currently live in the Arctic. There are always nomadic peoples, but more than two-thirds of this population lives in towns of more than 5,000 inhabitants. Despite its small and dispersed population, the Arctic has become an important region from both a geopolitical and economic point of view. These communities, especially the native ones, are vulnerable to climate change, but they also have a capacity for adaptation to change in their environment. However, some native peoples consider such adaptation unacceptable vis-à-vis their traditions and cultures. Thus the conference of Inuits complained to the U.S. Senate because it estimates that climate change is infringing on the rights of the community and risks leading to a loss of identity and culture.

The inhabitants of the Arctic already have and will increasingly have more problems linked to infrastructures. Some coastal regions will have to face simultaneously the melting of the permafrost, an increase in the frequency and size of storms and thus of erosion, and a rise in sea level. In the work that the Argos Collective devoted to climate refugees, Guy-Pierre Chomette and Hélène David describe their stay in Shishmaref on the northwest coast of Alaska. They relate the story of a house being discovered one morning in the sea. It had been balancing for a long time on the ocean banks and finally could not resist the erosion caused by the melting of the permafrost. Most of the 600 inhabitants of Shishmaref voted to leave their village by 2015 but without a very clear notion of their futures. The least onerous solution would involve leaving everything for the little towns located 300 kilometers farther south. A closer relocation, at some twenty kilometers inland but twice as expensive, is also foreseen.

The Political and Economic Stakes: Climate and Oil

If the current melting rate of the Arctic ice shelf continues, this will enable new links between the Atlantic and the Pacific, intensifying commercial transport and access to new fishing zones. But more important in terms of political and economic stakes, the floor underneath the ice shelf seems to contain 10 to 50% of the planet's reserves of oil and gas. But to whom do these ocean floors belong? There are five coastal countries—the United States, Canada, Russia, Norway, and Denmark, through its sovereignty over Greenland—but the answer to the question depends on who has rights to the sea; a state bordering the Arctic Ocean can claim rights over the undersea territories located beyond 200 miles from its coast if it can demonstrate that they form a natural prolongation of its continental shelf. That right implies a scientific expertise based in particular on a detailed cartography of the floors. The Canadians are already financing new submarines to explore the ocean floor; the Russians have used a bathyscaphe to plant a flag of the "motherland" in rustproof titanium on the ocean floor of the North Pole. Some non-coastal countries propose free access to the waters that they consider international. Fortunately the scientific projects within the framework of the 2007–2009 International Polar Year (IPY), thus in that of an international collaboration, lean toward ensuring that scientific interests supersede economic and political ones. It would then be a matter of perfecting technologies that would enable the energy treasures in the Arctic Ocean to be exploited.

As far as Antarctica is concerned, potential energy reserves are now under lock and key as a result of the Madrid Protocol signed in 1991 and in effect for fifty years unless the pressure of energy demands cause the signatories of the treaty to revisit it.

The issue of climate warming has caused scientific challenges and those that concern society to become closely connected; we can see proof of this in the high-stakes polar regions in a world needing energy resources. The polar zones seem to be peaceful spaces in an agitated world. Let's hope that the challenges connected to climate warming and to the race for resources will safeguard those spaces of peace.

CHAPTER 15

What We Must Do

We often have the feeling that there is an absence of dialogue, an uncrossable chasm, between the scientific world and that of the political policymakers. In the case of climate warming associated with human activities, the fact that the IPCC, which is responsible for scientific assessments, was founded by two organizations that came out of the United Nations has largely facilitated the dialogue. Four years after the creation of the IPCC, in June 1992 during the first Earth Summit organized in Rio under the aegis of the United Nations, 156 countries adopted the text of the United Nations Framework Convention on Climate Change (UNFCCC), known after 1994 as the Climate Convention. Two other conventions were then held in Rio, one dedicated to the preservation of biodiversity, the other to the struggle against desertification.

The Climate Convention has been ratified by almost every country on Earth (189 governments including the European Community) and is filled with much good sense because the ultimate objective of the convention is "stabilization of greenhouse gas concentrations in the atmosphere at a level that would prevent dangerous anthropogenic interference with the climate system. Such a level should be achieved within a time frame sufficient to allow ecosystems to adapt naturally to climate change, to ensure that food production is not threatened and to enable economic development to proceed in a sustainable manner."[1]

But that good sense, which consists of taking action because we cannot indefinitely allow the warming of our atmosphere to increase, creates a true challenge, even if the Climate Convention does not fix the level of stabilization that should not be surpassed.

Stabilizing the Greenhouse Effect: A True Challenge

We have pointed this out clearly: none of the scenarios cited in chapter 13 leads to a stabilization of the greenhouse effect. It is logical, since the mandate given to economists when they were asked to propose that set of scenarios was that they do not take into account policies put into play explicitly to fight against climate change. On the whole, none of those scenarios corresponds to what must be done, at least from the perspective of the Climate Convention. The recognized need to stabilize the greenhouse effect has given birth to a second type of scenario, that of stabilization (Table 15.1).

An initial approach, followed in the second and third reports of the IPCC, consists of defining the objective of stabilization in terms of CO_2 concentration because, as we have emphasized, it is unrealistic in the long term to seek to stabilize the greenhouse effect if the concentration of CO_2 is not stabilized. We must add approximately 20–30% to get a CO_2 equivalent that then accounts for all the greenhouse gases. We need only examine the constraints posed by the stabilization of concentrations of CO_2 alone to immediately grasp the challenge we face: this objective implies that the emissions of this gas are compensated for by sinks, regardless of the level of stabilization aimed for. If we are content with a stabilization of 1,000 ppmv of CO_2, 3.6 times more than the value in 1750, which from the point of view of a climatologist is completely unacceptable, it would be necessary for annual emissions never to exceed 15 gtC (billions of tons of carbon); it would take two centuries for emissions to go below their current level (~7 gtC). If we assign ourselves a more acceptable objective, 450 ppmv, these emissions should diminish by around 2020, reach their current value before 2050, and reach less than 3 gtC at the end of the century. Between the 30 gtC of the maximum scenario, the one in which no effort is made to limit the greenhouse effect, and that of a stabilization at a concentration not too far from its current value, there must be a decrease by a factor of 10 between now and the end of the century. The gap is huge and it would be useless to hope that nature could be a big help to us because the sinks, whether continental or oceanic, will have a tendency to decrease, at least in a relative sense. We must also not forget that in the long term the absorption of emissions that will persist will probably not exceed 0.2 gtC per year, less than 3% of the current emissions connected to fossil fuel—that is to say, nothing.

Table 15.1

IPCC stabilization scenarios.

Category	CO_2 concentration at stabilization (2005 = 379 ppmv)	Peak year for CO_2 emission	Changes in global, CO_2 emissions in 2050 (percent of 2000 emissions)	Global average temperature increase above pre-industrial at equilibrium, using "best estimate" climate sensitivity
I	350–400	2000–2015	$(-85) - (-50)$	2.0–2.4
II	400–440	2000–2020	$(-60) - (-30)$	2.4–2.8
III	440–485	2010–2030	$(-30) - (+5)$	2.8–3.2
IV	485–570	2020–2060	$(+10) - (+60)$	3.2–4.0
V	570–660	2050–2080	$(+25) - (+85)$	4.0–4.9
VI	660–790	2060–2090	$(+90) - (+140)$	4.9–6.1

Note: The six stabilization scenarios from the IPCC are displayed in increasing order as a function of the level of stabilization of CO_2 (column 1); the next columns indicate the years of peak emissions, the corresponding CO_2 emissions (compared to 1990), and the temperature of stabilization compared to its preindustrial levels (IPCC, 2007).

An alternative approach, adopted in the fourth IPCC report, consists of defining a mean temperature of stabilization that should never be surpassed. It has the advantage of enabling a relationship to be established between a given stabilization scenario and the impacts of climate change associated with it. These impacts (consequences) depend above all on the warming achieved, whatever the level of warming that has been reached within the context of a scenario of stabilization or not. It is enough to refer to the preceding chapter to have an initial evaluation of the impacts linked to a given stabilization scenario. The disadvantage of this approach is that it requires a hypothesis on the sensitivity of the climate vis-à-vis an increase in the greenhouse effect. The figures given in table 15.1 are based on a sensitivity of 3°C for a doubling of the concentration of CO_2, a value that coincides with the best recent estimates of a sensitivity of 2–4.5°C. Since it is important to have a few figures in mind, we have created a simplified table from the one presented in the IPCC report.

A few comments will be useful to facilitate reading the table. The six scenarios are classified in increasing order of the level of stabilization. For each scenario a certain number of simulations have been proposed; 70% of the cases studied for a given scenario have their maximum value of emission in the interval reported under "peak year," 15% above, and 15% below. Furthermore, the CO_2 emissions in 2000 were greater than those in 1990, a year of reference for the Kyoto Protocol, and the preindustrial temperature is lower by around 0.5°C than that of the period 1980–99, corresponding to what we designate, a bit arbitrarily, as the current climate. Finally, it is specified that emissions must continue to decrease after 2050. Table 15.1 illustrates the amplitude of the climatic challenge. Thus limiting warming to less than 2°C compared to the preindustrial climate may already seem beyond reach and that implies that we have to adapt to such warming, whatever we do otherwise.

This difficulty, if not impossibility, of stabilizing global warming below 2°C with respect to preindustrial climate is now clearly pointed out by specialists of energy. In its 2011 report titled "World Energy Outlook," the International Energy Agency (IEA) concludes that "We cannot afford to delay further action to tackle climate change if the long-term target of limiting the global average temperature increase to 2°C, as analyzed in the 450 Scenario, is to be achieved at reasonable cost. In the New Policies Scenario, the world is on a trajectory that results in a level of emissions consistent with a long-term average temperature increase of more than 3.5°C. Without these new policies, we are on an even more dangerous track, for a temperature increase of 6°C or more." The report adds that "If stringent new action is not forthcoming by 2017, the energy-related infrastructure then in place will generate all the CO_2 emissions allowed in the 450 Scenario up to 2035, leaving no room for additional power plants, factories and other infrastructure unless they are zero-carbon, which would be extremely costly."

The Kyoto Protocol: A First Step

Let's look back a bit. Once the Climate Convention was implemented in 1994, the signatories designated by the parties involved got to work. The first Conference of Parties (COP) gathered in Berlin in 1995. It recognized the privileged ties between the Climate Convention and the IPCC, which were solidified by the creation, within the convention, of the Subsidiary Body for Scientific and Technological Advice (SBSTA), whose role was to sum up for that body all the most recent information on climate changes including that provided by the IPCC.

That first COP included the Berlin mandate that aimed to reinforce the involvement of countries designated as Annex I in relation to those of the convention. This mandate stipulated, in particular, that the Annex I countries must assign themselves numeric limits and reduction objectives within precise time frames for their anthropogenic emissions and their greenhouse gas sinks. Annex I countries included relatively rich industrialized nations that had contributed historically to climate change, as well as countries in Eastern Europe and Russia that were "in transition towards a market economy" and had more flexibility in executing their obligations. All other countries, that is, developing countries and emerging countries, were in the non–Annex I group. These countries had to produce a more general report on their intentions vis-à-vis emissions and adaptations to climate changes. This mandate paved the way for the establishment of the Kyoto Protocol (COP 3) two years later.

COP 3 relied on the assessment presented in the IPCC's second report, which suggested a discernible influence of human activity on the climate. Even if that phrase is, rightly, very cautious, it has undeniably played a role in the decision taken by the Annex I countries to commit to numeric objectives and a calendar, since they agreed to reduce the carbon equivalent of their emissions by at least 5% over the period 2008–12 compared to that of 1990. The rate of reduction was different for each country: −7% for the United States, maintaining the 1990 level for Russia, +8% for Australia. As for the European Community, the effort of −8% required was, at the beginning, identical for each country. But after an internal discussion it was decided to modulate the effort in Europe country by country. For instance, Germany, which had just been reunified and could a priori easily reduce emissions of the former East Germany, was credited with a reduction rate of −21%. France, which emitted relatively less in the European context because a large part of its electricity is of nuclear origin, was required only to maintain its emissions.

The Kyoto Protocol specified that the Annex I countries first had to reach their objective through domestic efforts. But three mechanisms of flexibility offering other possibilities were also put into place. The first involved emission trading, which initially was controversial but later more widely accepted; a European system of emission trading was established in 2005. The second involved the Clean Development Mechanism (CDM), which enabled an Annex I country to use emission reductions achieved thanks to a certified project led in a non–Annex I country and leading to such reductions. Third, joint implementation was a mechanism to finance projects among Annex I countries.

The protocol was a new type of agreement on an international level, and the smallest details of its implementation were discussed in successive COPs; they were finalized in 2001 in Marrakech during COP 7. But it was still necessary for an essential condition of the protocol to be upheld, that is, that it be ratified by fifty-five countries representing at least 55% of worldwide emissions. The first element posed no problem because the non–Annex I countries, which were signatories but without numeric objectives, ratified it quickly. It was more difficult for the second requirement to be met because the Bush administration refused to propose ratification by the United States, and as long as Russia and a few other countries were not involved, the required 55% was not reached. It was reached in November 2004 when, on the initiative of Vladimir Putin, the Russian Duma became involved. We should note that this country, rich in emission credits (we speak in this case of "hot air"), had a certain interest from a purely economic point of view. Three months later, on February 16, 2005, the Kyoto Protocol was in effect.

Early on Australia followed the United States, at least until November 24, 2007, a symbolic date in the context of the struggle against climate warming. Kevin Rudd, leader of the workers' party, then became prime minister, winning the elections that pitted him against John Howard after an electoral campaign based largely on a pledge to ratify the Kyoto Protocol, which the latter obstinately refused to do. The following December 12 during the Bali COP Kevin Rudd confirmed Australia's pledge and received a standing ovation by an audience completely committed to the cause. By contrast, Canadian prime minister Stephen Harper, also elected in 2007, announced his intention to take Canada, a signatory country, out of the protocol. This struggle against warming had undeniably entered fully into the political arena; this could be seen in the United States where many cities and states expressed their disapproval of the Bush administration's position.

The last available emission inventories are for 2005. It is very unlikely that the decrease objectives pledged by the developed countries will be reached. The apparently "good students" (those that met their goals) are Russia (close to −30%) and the former countries of the East, such as Romania (−46%). They have benefited from favorable ceilings and from eased improvements at the level of energy efficiency. They also have been in the midst of an economic recession. Large European countries (Germany, England, France) are on the right path (France has reduced its emissions by 1.6%, which it aims to maintain), but others, in particular the countries to the south (Portugal, Italy and Spain) are very behind: Spain has increased its emissions by 53% whereas its objective was to increase by only 15%. Canada was also among the "naughty" countries. As a result, the Annex I signatories of the protocol (which had achieved a reduction of 15%) are collectively on the path of fulfilling their overall objective of −5%, largely thanks to the countries on a transitional path, whereas the European Community taken as a whole is in the red, having diminished by only 1.5% whereas it aimed for an 8% reduction. The situation looks less encouraging if we take into account Australia, which was not a signatory in 2005, and the United States, two countries that have increased their emissions by 26% and 16%, respectively; taken as a whole, the Annex I countries have nevertheless decreased their emissions by around 3%.

The Kyoto Protocol was only a very first step for two reasons: the modest objective for the developed countries and the extremely rapid increase in emissions of the non–Annex I countries, particularly those of the large emerging countries such as India, Brazil, and China, whose CO_2 emissions have multiplied respectively by 2.02, 1.65, and 2.38. If we combine all the greenhouse gases, we note a large increase in emissions between 1990 and 2004—24% on the scale of the entire planet. This increase has continued at an impressive rate. Over the period 2000–2008 CO_2 emissions have increased at an annual rate of 3.4% whereas that rate was on average 1% during the 1990s (data provided by the Global Carbon Project including emissions from fossil fuel combustion, cement production, and land use change).[2] Instead 2009 has been marked by a decrease of these CO_2 emissions (by −1.2%), but this was due to the economic crisis, and emissions increased rapidly the following year with a jump of 5.2% between 2009 and 2010 with total emissions reaching 10 GtC (to get gigatons of CO_2 this figure should be multiplied by 3.67). The distribution of these emissions has also greatly changed. Those of the developed countries, which in 1990 had around 60%, no longer represent more than 45%, and the resulting 55% are attributed to emerging countries whose emissions are steadily rising and to developing countries. But it is useful here to recall that these figures would be inversed if the emissions were attributed to the countries who consume goods and not those that produce them. With this type of calculation, whose logic is understandable, the developed countries would be seen as contributing around 55% of all emissions.

We are convinced that Kyoto was an extremely important step, despite what some people such as the "skeptical ecologist" Bjorn Lomborg think. He asserts that the efforts demanded are useless since in 2100 they will have decreased warming by only 0.15% and will slow it down only by six years.[3] Lomborg actually engaged in some sleight-of-hand by relying on a climatic simulation in which two scenarios of carbon dioxide emission are used. In the one used for reference, the annual use of carbon goes from its current value of close to 7 gtC to 20.6 gtC at the end of the century. In the second, the Kyoto agreements are upheld between 2008 and 2012, then emissions continue at the same rate as for the scenario of reference, with 2100 emissions that reach 19.5 gtC. In plain language, Lomborg is right, but he is tricking the reader. Kyoto alone followed by a resumption of emissions at their earlier rates accomplishes nothing or almost nothing. What is necessary in order to be climatically efficient is an ambitious post-Kyoto scenario that puts us on the path to a stabilization of global emissions by around 2015—but their decrease also has to continue.

The Bali Conference

Bali, Saturday, December 15, 2007: the closing session of the thirteenth Conference of Parties, the seventh in which one of us, Jean Jouzel, participated as part of the French delegation. This COP was probably the most tense, even though the Nobel Peace Prize had been awarded the preceding Monday to Rajendra Pachauri as an IPCC collective co-laureat and to Al Gore for their involvement in the struggle against climate warming. Both of them then came to Bali, and Al Gore delivered a speech, brilliant as usual but also extremely virulent against the Bush administration. It was not a last-ditch effort by the COP but the beginning of a process that by 2009 would be made concrete by an agreement that would follow the Kyoto Protocol. The ambiance was charged; early on, the United States, represented by Undersecretary of State Paula Drobriansky, refused to adhere to the decisive project written after long negotiations and then remained alone, completely alone, without their traditional allies in such circumstances: Australia, Japan, and Canada. They then had to make the best of a bad situation. The decision of Bali, which everyone agreed was *a minima*, was adopted.

Let's return to the major issues raised at this conference: for a "post-Kyoto" agreement to be operational by January 2013, everyone agreed that it should be set by 2009 so that it could be ratified by the greatest number of governments between 2010 and 2012. At the COP of Montreal in 2005, everyone was aware of that target date, but some countries, primarily the United States, did not want those negotiations to take place. An escape hatch had nevertheless been identified: it consisted of putting into place a special working group responsible for establishing a dialogue between the different parties involved that would present its report on the possible paths to negotiation two years later in Bali. That working group functioned well and reached some conclusions relying largely on the work of the IPCC—that is, that global emissions of greenhouse gases must reach their peak in the next ten to fifteen years, then by 2050 be reduced to levels located well below half those of 2000. It recognized that to achieve the lowest objective of stabilization it would be necessary that the developed countries decrease their emissions by 25–40% by 2020 relative to those of 1990.

Discussion focuses on the inclusion, or not, of these numeric data in the Bali proposal. The United States, which was joined by Canada, Japan, and Russia, was firmly opposed. According to the Americans, those levels of reduction were indeed what had to be discussed by 2009. As for Europe, it wanted everything possible done to ensure that warming did not exceed 2°C relative to the preindustrial period. And the developing countries, which rightly saw a self-serving policy by the developed countries as likely to lighten their potential involvement as much as possible, were favorable to it. Furthermore, the United States wanted involvement on the part of the latter countries. The United States wanted the involvement of the developing countries to be formulated in a similar way to that of developed countries, although the developing countries were of course different. And since we had to reach a consensus, each one took a step. The numeric data disappeared from the text, but it was referenced in notes at the bottom of the pages in the IPCC conclusions. We must admit, it was rather ambiguous, but it nevertheless favored scenarios I, II, and III (Table 15.1), the least emitters, as a future basis for discussion. The developing countries then joined the game by accepting the idea that they could put into place actions aimed at limiting their emissions within the framework of durable development in a measurable, reportable, and verifiable way. Those specifications would also relate to transfers of technologies and to financial assistance that the developing countries expected for their part with a view toward facilitating those actions. Of course the French and European negotiators were a bit disappointed, but they were also satisfied that a post-Kyoto agreement henceforth appeared within reach—or almost within reach, since all the countries, including the United States, ratified the fact that it should be finalized during the COP 15 that was to be held two years later in Copenhagen. Nevertheless, everything remained to be negotiated concerning the objectives in sight.

Can the Challenge Be Met?

To reach an agreement for limiting emissions beyond 2012 would be a great success; everyone is aware of this. This is especially so if the agreement is ambitious enough to allow for this indispensable division by two of the emissions, at least by 2050 with respect to their 1990 reference level. And it still must be achievable, technologically, economically, and socially. We will not describe here in detail the solutions that could be put into place, but we rely on the conclusions of the IPCC to convince the reader that it is within the realm of possibility.

According to the IPCC we can reduce the global emission of greenhouse gases during the next decades, and this could compensate for the projected increase in emissions or reduce them below their current level. Those possibilities must be researched in all sectors: energy production, transportation, construction, agriculture, forest management, and waste. In a given sector, no technology alone can provide the solution, but improving energy efficiency and eliminating energy waste are the greatest potential sources of reduction. The building sector can contribute in a huge way and benefit economically. Agricultural practices can also, for a minor cost, provide a significant contribution to the increase of carbon sequestration in the soil as well as a reduction of emissions and the production of biomass for energy use. Renewable energy should play an increasing role, such as the geological storage of CO_2 after it is trapped at the exit of thermal centrals using fossil fuels. Nuclear energy can also be used to reduce emissions in the electrical sector. Finally, transportation offers many options, even if they take longer to implement due to the inertia of our transportation systems and to constraints linked to the organization of our societies.

These possibilities can only be made concrete if the necessary policies are put in place. There is a wide range of measures available to governments, including taxation and regulation and voluntary agreements, as well as exchangeable emission credits and the clean mechanism projects mentioned above. The solution must thus rely on international collaboration but can also benefit from changes in lifestyles and individual behavior.

The IPCC stresses that all levels of stabilization can be achieved with already available technologies and those that will be commercialized during the next decades. The cost of this on a global scale is far from prohibitive. That cost increases in general as the level of stabilization aimed for decreases. It reaches its maximum value for scenarios I and II with, in 2050, a projected lowering of 5.5% of global GDP or a decrease in its annual rate of progression at scarcely higher than 0.1%.

The IPCC's message is clear: stabilizing the greenhouse effect at such an ambitious level is technologically as well as economically possible—only if, however, everything and everyone work together toward that goal.

From a more general point of view, it is very likely that the impacts of climate change will lead to increased annual costs as the global temperature increases. The IPCC analysis joins that of the Stern report,[4] which states that to master greenhouse gas emissions is economically more profitable than to allow them to grow without taking any measures to decrease them. Along the same lines the IEA's 2011 report, "World Energy Outlook," notes that "Delaying action is a false economy: for every $1 of investment avoided in the power sector before 2020 an additional $4.3 would need to be spent after 2020 to compensate for the increased emissions."

Copenhagen: Failure or Half-Success

The 2009 Copenhagen conference was a cruel disappointment for many of those who participated in or followed it. At best it was a half-success for those (in the minority) who wanted to remain optimistic. A few months after the conference, which was marked by the presence of around 120 heads of state or government, its outcome was viewed as ambiguous. The relative optimism that had resulted from the meeting in Bali was replaced, in 2010, with moroseness, and the conference was marked with pessimism that was accentuated by the attacks made against the fourth IPCC report and by the impact, largely due to media publicity, of theses developed by the "climate skeptics" for the general public.

The unusually high level of participation by policymakers reveals at least an awareness of the problems connected to climate warming. It is to Copenhagen's credit that there were discussions on the sources of innovative financing and on deforestation and biodiversity. That the Copenhagen agreement mentioned a warming limit of 2°C, which was firmly based on the IPCC report, was also a good sign. It also brought countries such as the United States into the struggle against climate warming that had excluded themselves or, like the large emerging countries, had no constraints on their emissions within the framework of the Kyoto Protocol. That this agreement is nonconstraining but based on voluntary involvement is perhaps not as negative as many fear, insofar as the promises made orally in Copenhagen have since been widely confirmed within the Climate Convention.

Yet we are, as scientists involved in research on the evolution of our climate, extremely disappointed in Copenhagen quite simply because the promises made are in obvious contradiction with the stated objectives. With 2020 on the horizon, we will have at best 5–10% more emissions in relation to the level required to have a good chance of limiting future warming to 2°C. The Copenhagen agreement was a cruel lack of ambition that should lead us to an average warming of at least 3°C by the end of the century. To illustrate the importance and above all the rapidity of this, we recall again that the passage from the Last Glacial Maximum (20,000 years ago) to the current climate, which was marked by an average warming of 5 to 6°C, took thousands of years.

At Copenhagen the large emerging countries made promises along the lines of the Bali commitments and even a bit beyond. The proposed plans are significant for China (40–45% reduction of emissions by point of GDP in 2020 compared with that of 2005), India (a similar commitment to that of China but at 20–25%), Brazil (decrease of 35–40% of emissions compared to the level foreseen for 2020), and Indonesia (decrease of 26% compared to the level foreseen for 2020). The only developed countries to commit to the Bali roadmap were Norway (decrease of 40% compared to the 1990 level) and Japan (total decrease of 25%). But this was not the case for the United States (decrease of 17% compared to the 2005 level or a decrease of 3% compared to the 1990 level, to be confirmed by legislative decision), or for Europe, which remained committed to a decrease of 20%, whereas a voluntary position taken on an objective at 30% would perhaps have helped unblock negotiations.

We strongly doubt that the revision of the Copenhagen agreement anticipated in 2015 will enable the objective of a warming limited to 2°C compared to the preindustrial climate to be upheld. This agreement was definitely adopted the following year in Cancun, but the 2011 Climate Conference held in Durban was again extremely disappointing. The second phase of the Kyoto Protocol, scheduled to start in 2013, will be launched but has been endorsed by only Europe, Australia, and a few other countries (representing about 15% of the global emissions). On the positive side, the idea of a platform to be completed in 2015 was launched at Durban with the goal of beginning in 2020 that would encompass 100% of the world's emissions. It has also been decided to trigger a process to close the gap between emissions anticipated in 2020 and emissions compatible with the 2°C target. This process will be made in light of the conclusions of the fifth IPCC report, which, for scientific aspects (Working Group I), will be published in 2013; the reports by the two other groups and the synthesis report will appear in 2014. Regarding the scientific approach that concerns us most directly, no article that has appeared since the publication of the fourth report seems to challenge its conclusions, which foresee that there will be an unavoidable warming that will become increasingly difficult (if not impossible) to control if we delay the implementation of an ambitious policy first of stabilization and then of decrease in greenhouse gas emissions on a global scale.

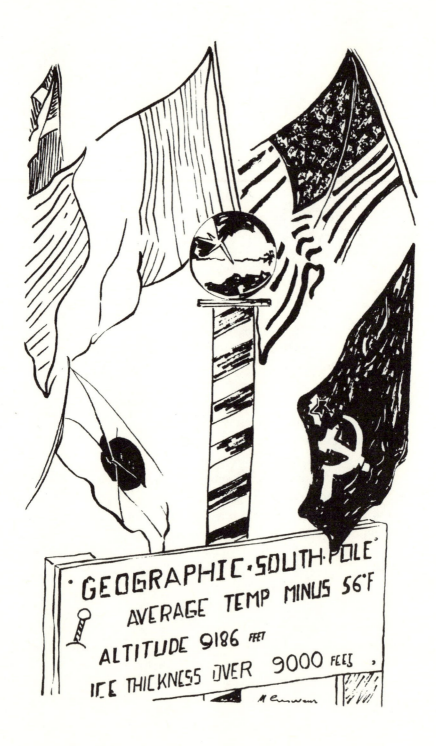

A Necessary Adaptation

The control of emissions must thus necessarily be accompanied by measures for adapting to climate change, which it seems impossible to avoid. Societies have long experience with managing impacts connected to meteorological or climatic events, but they will face a new situation in the coming decades. What is more, vulnerability to climate change can be exacerbated by other challenges resulting, for example, from current climatic anomalies: poverty and unequal access to resources, food uncertainty, globalization, conflicts, or certain illnesses. It is clear that the ability to adapt is strictly connected to socioeconomic development and that it is unequally distributed among societies and within them; but even societies that have a high capacity for adaptation remain vulnerable to climate change, to the variability of the climate, and to extreme climatic events. In fact, many options for adaptation are available in different sectors: water resources, agricultural practices, infrastructures and habitat (in particular in the coastal regions), human health, tourism, and transportation.

France has a national strategy for adaptation to climate change, one developed within the framework of a large commission led by the Observatoire National sur les Effets du Réchauffement Climatique (ONERC), which implicates various sectors of activity and civil society.[5] This strategy is organized around four great goals: to act in the interest of public security and health; to reduce inequality before risks; to limit the costs and derive potential benefits; and to preserve the natural patrimony. More than forty recommendations have been formulated that touch on the development of knowledge and observation, on information and sensitization, on control, and on mechanisms of financing. They also concern the prevention of risks of which a large number must be reexamined, as well as water resources and various sectors already cited: health, biodiversity, agriculture, energy and industry, transportation, building and habitats, tourism, banking, and insurance. Finally, they concern various geographic realms: the city, coast and sea, mountains and forests.

Neither on the global level nor on European or national levels does adaptation constitute a panacea for climate warming; it must accompany measures to improve and not to substitute. Many inhabitants of the planet, in particular those living in regions with temperate climates, can selfishly say to themselves that by 2050 they will be able to confront climate warming using only adaptation measures. This would be doubly selfish since by the end of the century adaptation will be increasingly difficult, even impossible, and not just for future generations but for young people who are already alive; as we have seen, a certain number of regions will be henceforth vulnerable to climate change. Fortunately that selfishness does not seem to be common and France, and all of Europe, proposes, at least in writing, an ambitious policy aimed at a reduction of greenhouse gas emissions.

The "Grenelle de l'environnement"

In France since the beginning of the 1990s, many reports examining the various aspects of climate warming have been written by the Académie des sciences, by various government committees, by various associations or interest groups, or following parliamentary initiatives; other reports dealing more specifically with energy explicitly take into account the climatic dimension. Moreover, the law of July 13, 2005, established the direction of France's energy policies and defined its strategy in the battle against an increase in the greenhouse effect: "The struggle against climate change is a priority of the energy policy which aims to diminish by 3% per year on average the emission of greenhouse gases in France. Consequently, the State is developing a 'climate plan,' updated every two years, presenting all the national actions in place to fight against climate change. Furthermore, in that fight, which must be conducted by all the States, France supports the definition of an objective of dividing by two the global emissions of greenhouse gases by 2050, which necessitates taking into account the differences in consumption among countries, a division by four or five of those emissions for developed countries."

This is the Factor 4 concept. In France it was introduced by Jean-Pierre Raffarin, prime minister at the time, during the opening session of a plenary meeting of the IPCC held in Paris on February 19, 2003, and has since then been discussed in various reports.[6]

The objectives of the Grenelle de l'environnement, the result of a commitment undertaken by Nicolas Sarkozy with ecologist associations, are much broader since they aim to take into account all environmental issues. The initiative was launched on May 18, 2007, by Alain Juppé, then minister of Ecology, Development, and Durable Management (Médad); Jean-Louis Borloo assumed responsibility for it when he succeeded him to that post as state minister. The name "Grenelle" harkens back to the Grenelle social agreements of May 1968, whose format that of the environment more or less adopted in aiming to gather all those involved around one table. In this spirit, each working group was composed of eight members from five groups: the state, local groups, employees, professional leaders, and nongovernmental environmentalist organizations and professionals. Six discussion groups were initially put in place, four dedicated to the great environmental themes— climate-energy, biodiversity and natural resources, health and environment, durable agriculture—and two cross-groups, one dealing with institutions and governance, the other with ecological development and competition. Then two subgroups were added, one devoted to genetically modified organisms and the other to waste.

One of us, Jean Jouzel, was called upon to preside, jointly with the economist Nicholas Stern, over the first working group called "Fight against climate changes and master the demand for energy."[7] Seen from the inside, we can testify to the excellent mind-sets and the quality of the contributions and discussion—as well as to the obvious enthusiasm, which was the mark of these meetings organized from July to September 2007 with a view toward developing proposals. Those involved locally (through regional meetings) and the general public (through the Internet) were then able to react to the measures proposed. Then those measures were discussed during a round table on October 24 and 25, 2007, and whose conclusions were then drawn at the Élysée by Nicolas Sarkozy, who committed himself to ensure "that all the specific, concrete, and consensual conclusions can be put to work."

This working group very quickly assumed the Factor 4 objective and adopted the intermediary stages set for 2020 by the European Council: reducing greenhouse gas emissions by 20%, or 30% in the case of what was promised by other industrialized countries; lowering the consumption of energy by 20%; and using renewable energy for 20% of the total energy used. The discussions were not always easy, but they were constructive enough so that we developed a set of measures which, if they are put into place in a voluntary and effective way, will be able to put France on the trajectory defined on a European level and then, we strongly hope, on that of Factor 4. All these measures[8] were agreed upon by the majority of members of the group, even if some have been met with reservations or opposition; the only point of extreme disagreement was the future of nuclear plants.

More than thirty operational committees were established to prepare legislative documents that were examined by legislators in the law called Grenelle I, which was voted on in 2008. A second law, Grenelle 2, voted on in 2010, concerns more directly the application of measures to be put into place and part of them already put into effect. We must note that these only partially represent the ambition of the proposals made in 2007. But we retain the hope that France will continue to be a driving force in the future in the preservation of our environment, and more specifically in the fight against the greenhouse effect. Just after his election in 2012, the new president, François Hollande, launched a national debate on energy ("La transition Energétique") and plans to reactivate the Grenelle de l'environnement under a slightly different format. Let's keep our fingers crossed.

PART FOUR

THE POLES AND THE PLANET

CHAPTER 16

The Crucial Place of Research

In this book we have traveled the white planet, the one formed by the ice, which for the most part is found in the polar regions. We have gone deep within the inland ice sheets, in Antarctica and Greenland, to discover the wealth of the glacial archives. The warning sign we have been given cannot leave us indifferent: in this new era, the Anthropocene, humans mark the environment of their planet with their imprint and above all the climate in which they live. Our message aims to help convince citizens and policymakers of the urgency of the measures that must be taken to respond to the challenge of climate warming and the degradation of our environment because we must act and not simply endure. In immediate terms, the scientists of the International Polar Year have listened to the ice, including the fragile shelves, which has an impact on our near and more distant future. We will first talk about research before describing the rise in pollution, then we will look at the connections between the poles, the planet, and our societies.

We will return to our white planet by discussing a theme that connects them, that of the growing influence of research. We will recall the importance of the contribution of fundamental research, essential to an understanding of the mechanisms that rule our climate, to predict its evolution and to provide a correct evaluation of the uncertainties associated with it. The quality of this fundamental research and the effort led by the scientific community to construct, through the IPCC reports, an easily accessible collective assessment have, moreover, constituted a point of departure for the process undertaken to fight against climate warming.

We would like to point to the multidisciplinary nature of the research concerned with climate change around which mathematicians, physicists, chemists, and geochemists, and specialists in the biosphere and in past climates, come together; when we see the impacts of the warming to come, this should also mobilize ecologists and specialists in biodiversity, public health, and agriculture, as well as researchers in the human and social sciences and economists.

We should recognize that the objective of decreasing greenhouse gas emissions and mastering energy constitutes a true challenge, whether on the horizon of 2020 or 2050. Climate specialists can contribute to that goal by reducing the uncertainties surrounding the evolution of extremes and climatic variability, the carbon cycle, the sensitivity of the climate, the risks of surprises, the regional characteristics of climate change, and the different impacts that will be associated with them. Specialists in the human and social sciences, economists, and legal experts will play a key role in implementing measures aimed on the one hand at stabilizing the greenhouse effect at an acceptable level and on the other at adapting to the climate warming with which we are faced. Added to this, of course, is the need to develop the technology necessary to reach the set objectives in terms of greenhouse gas emissions and energy mastery throughout all the implicated sectors: transportation, energy, housing and urban planning, waste, agriculture, and sylviculture.

The scientific community involved in the study of snow and ice provides an increasingly important contribution to a knowledge of our climate, alongside specialists of the dynamics and of the chemistry of the atmosphere, oceanographers, and specialists in the biosphere. This contribution involves sea ice, glaciers, ice caps, large ice sheets, and even permafrost; our community is concerned with their past, present, and future evolution. The growing importance of this contribution is also well recognized on the international level because, since 2000, the World Research Program on the climate has made the study of the interactions between the climate and the cryosphere one of its beacon programs, the Climate and Cryosphere (CLIC). Furthermore, the evolution of our white planet is at the heart of a number of potential repercussions of climate change that affect sea level, water resources, geopolitical issues, and aspects tied to the fauna and flora, tourism, and even the future of populations and the economy in certain regions around the Arctic Ocean.

Beyond a better knowledge of the evolution of our climate and the various impacts of climate change, the polar regions offer data for many other areas of research, from the observation of the interior of the Earth or the cosmos to following the composition of the atmosphere and the way in which the ozone hole will be reduced in the coming decades. The research opportunities are enormous, and the entire community of polar scientists became strongly involved, between 2007 and 2009, to ensure the success of the Fourth International Polar Year (IPY).

A Short History of the Polar Years

This IPY follows three initiatives taken to explore and study the confines of the Earth. The first took place in 1882–83: twelve countries organized thirteen expeditions to the Arctic where they benefited from the data collected from fifteen observatories.

In 1932–33 the World Meteorological Organization was interested in the poles not only for climatology but also with regard to the magnetic field and northern lights. Forty countries participated in it; a network of stations were established in Arctic and sub-Antarctic regions.

The International Geophysical Year (IGY) 1957–58 was then proposed by physicists. When proposing the IGY, physicists had the idea to benefit from a peak in solar activity to study the connections likely to exist between the Sun and terrestrial phenomena, such as the magnetic field and northern lights, and between the Sun and the gases present in the upper atmosphere. The IGY used new techniques developed during World War II and benefited from the launch of the first satellites. That polar year was remarkable for several reasons. For the first time, under the patronage of the International Council of Scientific Unions (ICSU) and science academies, twelve countries set up forty-eight stations in Antarctica, four of which were in the interior of the continent. Observatories and fieldwork would enable scientists to study the atmosphere as well as the ice sheet. During the cold war the IGY was a huge scientific success and led governments to sign the Antarctic Treaty in 1959, which dedicated that continent to peace and research—a status that remains unique in a world that is always agitated by many conflicts.

That status owes a great deal to the involvement of scientists who, at the initiative of the ICSU, established the Scientific Committee for Antarctic Research (SCAR) in 1958. In the beginning this committee brought together twelve countries; it now includes forty and plays an essential role in the impressive development of research in Antarctica. The situation is different and more complex in the Arctic. Although each of the countries bordering the North Pole (Canada, Denmark, the United States, Finland,

Iceland, Norway, Russia, and Sweden) has over time set up observatories there, the conflict between Americans and Russians impacted international collaboration throughout the cold war. It was a speech by Gorbachev, which in 1987 led to the disarmament treaty, that finally opened the path to broad cooperation that a bit later was solidified by the creation of the Committee for Sciences in the Arctic; here, too, research was integral to a more political structure, the Arctic Council, which was created in 1996.

During the IGY the launching of the first satellites—Sputnik, then Explorer 1 and 3—enabled the discovery of the Van Allen Belt, as well as an understanding of the formation of southern and boreal lights, by demonstrating that the particles of solar wind are channeled in polar regions by force lines of the magnetic field. The impact of these high-energy particles with the gases of our atmosphere leads to the formation of high-altitude lights, between 100 and 1,000 kilometers high, visible on a clear night in the form of wonderful sheets. Our planet resides inside this shield.

After the first scientific fieldwork on the ice sheets (at the beginning of the 1950s in Greenland and during the IGY in Antarctica), we became aware of the volume of that ice whose melting or extension has caused the sea level to vary by up to 120 meters during the Quaternary. Repeating the long drifting of the Norwegian explorer Fridtjof Nansen, a voluntary prisoner of the ice shelf in the Arctic in 1895, stations drifting on the ice, then submarines, have enabled us to define the very large expanse and the thinness of that ice shelf. Considerable progress was then achieved with the use of satellites, which enabled us to inventory the "White Planet" and its evolution.

Measurements of the quantity of CO_2 in the atmosphere began during the IGY, but it was in the 1980s that the ice core drilling undertaken in the heart of Antarctica enabled us to show for the first time the connection, during the last few hundred thousand years, between the climate and the amount of greenhouse gas in the atmosphere. We were also able to prove the rapid and large rise of the greenhouse effect in the last two centuries, which the IPCC concluded was likely the major cause of the climate warming of the last few decades. The increase in the concentration of greenhouse gases in our atmosphere since the beginning of the industrial era is moreover only one of the aspects of the degradation of our environment on a planetary scale.

The use of new techniques has also been applied to life sciences. Thus after the discovery of a rookery of emperor penguins in Adélie Land at the beginning of the 1950s, ultraminiature signal receivers later enabled scientists to follow by satellite the emperor penguins in their long march over the ice shelf and their plunging into the water, and to study their collective behavior that ensures the survival of the group as well as that of individuals.

The International Polar Year 2007–2009

For the 2007–2009 IPY, the idea was to take the pulse of these regions and to study their role in the Earth's system. The focus was on the current problems of society in which the polar regions play an essential role (environment, climate, biodiversity, future of local civilizations, etc.). It also opened onto fundamental research in new horizons, going from the infinitely small (microbiology) to the infinitely distant (astronomy), which will feed the discoveries of the future.

We may of course be surprised by the need to concentrate research on an international level on a given geographic zone during a period that lasted only two years. The choice of the poles is tied to the importance of the ice there in one of the great current challenges of our society: climate warming. To better anticipate what will happen in the future, the two patrons, ICSU and WMO will simultaneously study the two polar regions during an annual cycle. Taking into account the hemispheric disconnect in the seasons and the difficulty in accessing the high latitudes in the winter, the IPY thus unfolded over two years, from March 2007 to March 2009. To be accepted, a program must lean on a large international participation, including China, the United States, Europe, South America, and Australia. In all, sixty-three countries combined their resources in researchers, observatories, oceanographic ships, ice-breakers, convoys over the ice sheets, helicopters, airplanes and satellites, and computers. More than two hundred projects were organized around major themes such as the study of climate warming and its impact on the ice, biodiversity, and populations in the Arctic, which are also subjected to great socioeconomic changes—so many societal concerns in a world in full evolution. The work of the scientists has been spread by many systems in order to inform citizens and policymakers and to attract a new generation of researchers.

As climate warming and its impacts are very marked in the polar regions, it is logical that many IPY programs involved the role of those regions in the global climatic system. In the Arctic, the increase in temperatures is two to three times more marked than on average around the globe and, as we have mentioned, many impacts are already visible: the degradation of infrastructures with the melting of the permafrost; the threatening of the resources of the local populations whose subsistence depends on fishing and hunting; and the threatening of species, such as the emblematic white bear. In Antarctica, following the retreat of the ice shelf around the peninsula, the observation of the Southern Ocean floor has, since the beginning of the new polar year, enabled scientists to discover ecosystems of exceptional wealth, including thirty or so new species.

Other programs involve more basic research. The "Ice Cube" project at the South Pole looks at the infinitely distant by using the perfect transparency of the deep ice to detect cosmic neutrinos from the weak bluish luminous trails they leave in their path and, from there, to attempt to determine their source in the universe. With this goal in mind, a network of detectors was implanted in a volume of ice from one cubic kilometer to a neighboring depth of two kilometers.

The French contribution to the IPY is very significant including, among other places, in the Arctic, with the study of glaciers, the ice shelf, and pollution, and in Antarctica with the resumption of exploratory fieldwork, new ice core drilling, and the development of scientific activities at Concordia Station in the heart of the ice sheet. We cannot discuss all of these projects here, but we would like to share the approach of and the hopes for three of them, in which we are involved in various ways. We will first mention the scientific objectives in the realm of glacial ice core drilling. Then we will discuss the research on traces of life in Lake Vostok, which is under nearly four kilometers of ice; even if, for the moment, no water has been extracted from the lake, plans to do so exist, and an analysis of the deepest ice of the Vostok coring formed by the refreezing of the subglacial lake water has already opened paths in which Dominique Raynaud has, with colleagues from Grenoble and Russia, been very interested during the last few years. Finally, beyond its undeniable aesthetics, the Franco-Italian Concordia station is an ideal base for research in numerous disciplines; two of us, Claude Lorius and Jean Jouzel, have had the privilege of following its creation, as presidents of the administrative council of the French polar institute. The station's genesis owes a great deal to the involvement of both groups' successive directors, Roger Gendrin and Gérard Jugie, and to Mario Zuchelli, their Italian counterpart.

Glacial Ice Coring: Ambitious Objectives

Each core drilling in the ice provides new elements for describing and understanding the evolution of the climate. Over the next decade, the main priority of the internationanl community of glaciologists is to deepen the inquiry from data of the past. This community proposed and created the International Partnership in Ice Core Sciences (IPICS) with four targeted projects.

The first is to obtain a complete record of the last interglacial period in Greenland or an ice core covering at least the last 140,000 years, which that of North GRIP was not able to provide, as it stopped before 125,000 years and revealed nothing of the conditions that prevailed in Greenland during the next-to-last deglaciation. The research of the optimal core site to the north of North GRIP began with the IPY in 2007. As already mentioned, this new international drilling project led by our Danish colleagues under the directorship of Dorthe Dahl-Jensen was completed during the summer of 2010. This project, for which Valérie Masson-Delmotte has taken responsibility vis-à-vis French participation, should enable a better understanding of the evolution of the ice sheet of Greenland during the preceding interglacial period; it is important because the climate, which was warmer than what we are currently experiencing, had caused a reduction in the volume of ice, but no one really knows in what proportion.

Going back further in time enriches our knowledge, as has been shown in the extension of the time scales between the recordings at Vostok and those of Dôme C. From this perspective, the second project consists of researching one or several drilling sites in Antarctica that will enable us to gather climate archives at least 1.2 million years old. Why that objective? Quite simply because marine sediments from that era reveal a transition between glaciations that were less intense and had a periodicity close to 40,000 years and more marked climate cycles occurring every 100,000 years. Only the glacial archives are able to describe the evolution of Antarctic temperatures, the composition of the atmosphere, and the correlation between climate and the carbon cycle during that true climatic revolution of the middle of the Pleistocene, whose origin is far from clear. The hypothesis of a connection between that major transition in the rhythm of glaciations and a long-term change in the atmospheric concentration of carbon dioxide could finally be validated. But first we must identify the sites likely to contain ice that is that old, a priori in regions where accumulation is the smallest possible and where the thickness of the ice is great and flow rate very low. It requires considerable logistical effort to explore the coldest and driest regions in the center of Antarctica. Scientific fieldwork, in particular by IPEV, going from the great drilling sites, Dôme C and Vostok, to the region of Dôme A, a priori the most promising sites (Figure 1.2), will be carried out in that spirit of exploration. A first shallow ice core has recently been taken by our Chinese colleagues.

To go back as far as possible in time, to break records, is fortunately not the only goal of glaciologists. Thus the last 40,000 years are the object of the third project and of the greatest attention because that period encompasses both the rapid and great variations (of which we are far from having solved all the mysteries) and the last deglaciation. This explains paleooclimatologists' great interest in a series of ice cores covering that period both in Antarctica and in Greenland. The major challenge will be to construct the most precise chronological framework possible by using the different markers present in ice, which can be used to such a series of ice cores, such as variations in the concentrations of greenhouse gases and dust, and then— this is essential, in particular vis-à-vis the uncertainties that concern the future rise in sea level—to understand the response of the large ice sheets to a warming as great as the one that accompanied the last deglaciation. Ice core drilling in the coastal regions of Antarctica can enable us to follow the evolution of the ice cap and its flow during that period. That of Talos Dome, which ended successfully at the end of 2007, fits perfectly within these objectives.

Finally, with the fourth project, glaciologists are interested in the last 2,000 years in view of reconstructing the climate on a very detailed temporal scale, in some cases year by year, but also for measuring the evolution of the ice mass in relation to the rise in sea level. Beyond Greenland and Antarctica, data of this type can be obtained from many glacial ice caps in Arctic regions and from high-altitude glaciers in low and mid-latitudes that are not affected by melting.

The Microbiology of Ice and Subglacial Lakes: Life in an Extreme Environment

We often say that biology will be the science of the twenty-first century. What if that also applied to the vast deserts of the white planet? The story begins in Paris in July 1955. Eleven nations were represented to prepare the International Geophysical Year 1957–58. The countries involved essentially were researching the coastal sites in Antarctica to establish their research stations there because they would be easier to access than more inland sites. France installed the Charcot base at 320 kilometers from the coast, but the two giants, the United States and the Soviet Union, with more ambition because they had more means, also proposed the South Pole. The United States, which had been the first to express interest in that emblematic site, established a site at the pole. The Soviet delegation, which had arrived late in Paris because at that time obtaining permission to travel to the West was not an easy task, was disappointed. And if we are to believe Igor Zotikov as he relates in his book,[1] it was perhaps a bit out of spite and to show their ambition that the Soviets chose the place that is farthest from any coast: the pole of inaccessibility, close to the geomagnetic pole. Ten enormous treaded Kharkovchanka vehicles left Mirny Station on the coast in the direction of the geomagnetic pole, which they reached on December 16, 1957. Vostok Station was born. That station has offered glaciologists 420,000 years of climate archives; a wise choice indeed.

Without the Soviet delegation's late arrival to the meeting in Paris, the discovery and the destiny of the largest subglacial lake in Antarctica would no doubt have been different. The Soviets established their Vostok base on the high plateau of East Antarctica without knowing that under their feet and under around four kilometers of ice was a huge lake with a surface equivalent to that of Lake Ontario. At the time theoreticians could already foresee that because of the geothermal flow emitted by the rock base, the temperature of the ice could likely reach the point of fusion under several kilometers of ice, despite the extreme temperatures that existed on the surface. But from that to imagining the existence of a gigantic lake some hundreds of meters deep—no one had made that leap.

In the 1970s subglacial lakes were discovered in Antarctica by a team from the Scott Polar Research Institute of Cambridge University directed by

Gordon Robin.[2] With the goal of studying the structure of the inland ice sheet, that team made a series of flights with American C130 airplanes using a new method of radiography of the ice sheet with radar equipment implanted in the plane; the radar emitted radio waves that penetrated into the ice and were reflected by the different layers encountered, including the rock base. Imagine the surprise of this team upon observing over some of the many radar images covering a large region of East Antarctica that the rock base, generally irregular, was interrupted by extremely flat zones. Because water reflects the radar waves of low frequency of a rock base more strongly, Gordon Robin attributed these horizontal echoes to the presence of subglacial lakes. During this period of intense aerial radiography of the Antarctic continent, many flights were taken in the Vostok region using the station to facilitate navigation. Many "flat echoes" were recorded, which led to the suggestion in 1977 of the presence of a huge subglacial lake in that region. That marked the discovery of Lake Vostok.

In 1991 the launch of the European Remote Sensing satellite 1 (ERS 1) opened a new chapter in the saga of Lake Vostok. The eye of the satellite was able to describe in detail the topography of the surface of the ice sheet, and manipulating the satellite images then caused the incline of the surface to appear in different colors, thus demonstrating the zones that had almost no slope. A zone of this type, of around 230 kilometers in length by 40 kilometers in width, was drawn at the place where the existence of Lake Vostok was indicated from the observation of radar waves. It was the imprint of the surface of the lake that lay under four kilometers of ice. The existence and the position of the lake were thus firmly established by the following evidence: the basal temperature rose under four kilometers of ice—confirmed by the measurements of temperature carried out in the core hole of Vostok—and flat topographies at the ice-lake interface. At the time definitive evidence was still lacking, that is, having a sample of the water from the lake. But since 1998 biologists who work in extreme conditions have had access to the ice formed from the water of Lake Vostok, the refreezing of which is found over a span of 80 meters in the deep part of the ice cores.

Biologists' interest grew as the story of Lake Vostok unfolded. The pioneers were indisputably the Russians, especially the Institute of Microbiology of the Academy of Sciences in Russia, but researchers' interest in the lake intensified when a part of the refrozen ice became accessible. Although it was difficult to extract data about the lake composition from this refrozen ice, that lake ice could provide precious information on subjects as fundamental as the development of primitive life in an extreme environment. Imagine life under 400 bars of pressure, around 0°C, and in the absence of solar energy and light! Could a living organism adapt to such conditions? The stakes go beyond the fascinating search for traces of life in those conditions when we consider the possible presence of a subglacial ocean covered with dozens of kilometers of ice on the moon Europa of the planet Jupiter. How can we not imagine Lake Vostok as a reduced, analogous model and even as a field of experimentation with a view toward the exploration of signs of life on other celestial bodies, like Mars or the Moon, on which scientists do not exclude the presence of water? This research could also be applied to the realm of genetics.

Within the framework of a collaborative agreement between the Russians, Americans, and French, they all have access to accreted ice samples from the lake. In the same issue of the journal *Science*, two international teams led by American researchers John Priscu[3] and David Karl[4] provide astonishing revelations demonstrating relatively high microbial concentrations in the accreted ice and conclude that Lake Vostok could contain viable microorganisms, even though it has been isolated from the atmosphere for more than a million years.

At the same time the Russians and French combined their efforts within the framework of a consortium that brought together glaciologists and microbiologists under the direction of Jean-Robert Petit from the LGGE and Sergey Bulat, microbiologist at the Saint Petersburg Institute of Nuclear Physics. They first demonstrated the need to use specific methods of decontamination, which had already been developed by glaciologists to measure certain trace elements in the ice cores. The application of these "ultraclean" techniques indicates that the density of microorganisms revealed in earlier publications is quite likely overestimated and that the accreted ice is a biological semi-desert. However, in 2004 a work by Sergy Bulat and colleagues held a large surprise by revealing the existence in that desert of a few thermophilic bacteria (which like the heat) whose closest "cousins" have been identified only in deep ditches or hot-water sources far from Antarctica.[5] It is difficult to suggest a contamination there; it is more likely that those bacteria prove the existence of hydrothermal sources in the deep crevices that could exist at the bottom of the lake, sources that have been proven thanks to the isotopic analysis of the helium trapped in the deep ice.

This was only the beginning of a long story and of a contribution that will certainly hold great surprises in biology, the science of the twenty-first century. The next step was to extract water from the lake. There was very strong interest among microbiologists in obtaining samples of that precious water because the accreted ice at the moment of its formation could have discriminated among the microorganisms that it accepted within its crystalline system and thus not be perfectly representative of the composition of the lake. But access to the lake raised a major problem, that of its contamination the moment it was penetrated, since the borehole was full of kerosene to prevent its closure over time. As described in chapter 6, the Russians penetrated the lake on the morning of February 6, 2012, likely with a minimal risk of contamination, as revealed by the water rise of about 600 meters above the lake surface in the drilling hole. Vladimir Lipenkov's words bear repeating: "Such was the end of the 5G drilling, but hopefully not the end of the Vostok project."

The Russian team plans to come back at Vostok next season (2012–13) and to start coring the 600 meters of refrozen lake ice in the hole. There is no doubt that a new step in the Vostok saga will be the discovery of the hidden world of Lake Vostok.

There are many subglacial lakes in Antarctica that are in general much smaller than Lake Vostok. So, why not try to test less risky penetration technology in one of these smaller lakes? Indeed there are new projects of penetration in such lakes, proposed by other nations (e.g., United Kingdom, United States). This is perhaps not as easy as it sounds because the limited amount of water in a small lake could be more sensitive to contamination. And above all, we now realize that there is a true subglacial hydrology that might connect several lakes.

The lake is only part of the cornucopia sleeping under the Vostok ice. Another part is what is formed by hundreds of meters of lake sediment accumulated throughout time at the bottom of the lake. This is an area of research for the next generation of scientists interested in the secrets of Antarctica.

Concordia: A Station Full of Promise

Who doesn't recognize those two elegant cylindrical buildings connected by a passage, lost in the middle of the huge Antarctic expanse, more than 1,100 kilometers from the French base Dumont d'Urville and 1,200 kilometers from the Italian base Terra Nova Bay? Made up of two main buildings three stories tall and having a total area of 1,500 m², the base was completed in 2005. The two buildings, constructed on stilts, were respectively dedicated to so-called calm (bedrooms, laboratories) and noisy (kitchen, restaurant, workshops) activities. The base can accommodate around fifteen people, scientific and technical personnel who must live in complete autonomy for nine months of the year; there is an annexed camp that can accommodate an additional forty people during the southern summer. The station can be reached only during a window of about ten days in the summer by ground convoy starting from Dumont d'Urville or by plane, generally from the Italian coastal base Mario Zucchelli of Terra Nova Bay.

The project took shape at the beginning of the 1990s. A Franco-Italian agreement was signed in 1993 between the polar institutes of those two countries with the objective of constructing a permanent Concordia Station at Dôme C. It was on this site that the European EPICA core drilling project, covering 800,000 years, was carried out during 1996–2005 and during the months of December and January. The deep ice core project, and that of the construction of the base, thus benefited from shared logistics largely provided by the IPEV, which, as heir to a tradition of several decades of French polar expeditions, became expert at getting material to the heart of the Antarctic continent. Thus the construction of the Franco-Italian Concordia Station and the completion of the deep ice core EPICA drilling necessitated the delivery of more than 4,000 tons of material at more than 1,100 kilometers from the coasts and at an altitude of 3,300 meters.

That heavy equipment was transported by ship from Hobart in Tasmania to Adélie Land, transferred over the continent via the ice shelf in winter, and then taken by land convoy within Antarctica. Each year three round-trips were undertaken during the southern summer between the coastal station of Dumont d'Urville and the Concordia site. Each convoy was made up of six or seven traction machines that towed some thirty hauling rigs, including trailers for personnel. All the vehicles and carts of the convoy were designed for use in Antarctica, whether they were tractors, platforms for cisterns, or containers. The average speed of the convoy was about ten kilometers per hour and a round-trip took twenty to twenty-five days. Technical improvements were continually made to increase the efficiency of the convoy while decreasing the impact on the environment (regulating the gases emitted from the motors and reducing the packing of the snow on the road).

An ideal place for glaciologists, Dôme C, which benefits from a particularly stable, pure, and dry atmosphere, was also perfect for studying the chemical composition of the low and high layers of the atmosphere. Researchers focused on the transfer of trace elements between the air and the snow, on the unleashing of katabatic winds, and on following the evolution and the reestablishment of the ozone layer. Because it is far from the coastal regions, Concordia is also ideal for observations in magnetism and seismology, while the isolated and confined living conditions in which a group of winter dwellers coexist are propitious for carrying out biomedical programs applicable to long space flights.

Astronomers and astrophysicists are also very interested in Dôme C. Thus the competition among astronomers worldwide to go ever further in the exploration of the universe's history and to obtain ever finer details of celestial objects leads to the development of new techniques and to the discovery of new earthly sites that can improve the performance of current technology. The advent of space tools opened a new dimension in the astronomical observation, as did the perfection of the adaptable optics in earthly observatories. These two systems are, however, onerous and have their own limitations in terms of both accessibility and technological progress. An alternative has appeared with the installation of permanent continental Antarctic stations located at high altitude on the plateau. Astronomers are very interested in using these exceptional sites for several reasons: there are long polar nights that at the center of Antarctica extend over several months; there is the conjunction of very low temperatures and the remarkable stability of the lower layers of the atmosphere; and there is an absence of anthropogenic perturbations. The exceptional quality of the site of Dôme C vis-à-vis these criteria has been acknowledged; the only potential competitor is the site of Dôme A, where a permanent station has not yet been built. Thanks to Concordia Station, two great research projects could be rapidly developed at Dôme C: the detection of planets similar to ours (exo-Earths) and that of traces of the first moments of our universe in the cosmic depths.

But it is time to return to Earth and even into the ice to discuss our planet.

CHAPTER 17

Humans and the Rise of Pollution

The degradation of our environment is one of the greatest challenges facing our society. With an increase in population from 1.6 billion in 1900 to 6.6 billion at the beginning of the twenty-first century, with all of its needs and activities, the impact of humans on the planet is now more than worrisome. That impact is manifest both on a local scale, very close to sources such as the great megacities of the industrialized countries, and on a global scale. To increase our awareness of this and to evaluate the state of the health of our environment, we can examine the polar ice caps, which, despite their apparent purity, contain rich data. In those desert and distant zones, they have recorded the rise of pollution and the global nature of the impacts of our activity, a story that should lead us to serious reflection. Let's look at a few aspects of this story.

A few centuries ago explorers and naturalists realized that the equator did not represent a border between the Northern and Southern hemispheres. Going to the high latitudes of the two ends of the planet, they observed the flight of the Arctic stern birds migrating with the seasons and sighted the great humpback whale as far as the great South, animals guided who knows how in their flight and their swim. These observations suggested that we have only one atmosphere and one ocean.

For the glaciologist of polar regions, the first unexpected discovery came from the observation of magnificent boreal and southern lights that over the course of a few months went from the nights of the ice sheet of Greenland to those of Antarctica—a phenomenon whose study dates from the International Geophysical Year fifty-five years ago: high-energy particles from the solar wind were channeled by lines of force from the magnetic field that surrounds the Earth, and it was at the poles that they encountered the gases of our atmosphere, creating colored sheets in high altitudes. That magnetic field surrounds the Earth, protects and isolates it, and it is in that shell that life was able to develop.

In the interior, during campaigns from one pole to the other, glaciologists have been able to discover traces of human activity. Although we live in it—or rather survive in it in all of our large cities—there are glacial archives that reveal the undeniable rise in the pollution of our atmosphere on a global scale since the advent of the industrial era, which began at the beginning of the nineteenth century, well before the installation of scientific measurement networks.

Aerosols, particles suspended in the atmosphere, and gases can be natural or a result of human activity. Some pollutants, gases with long lives, and small aerosols, or those that have reached the stratosphere, travel over great distances and meet up in the polar regions and in their ice. Since most of the industrialized countries are in the Northern Hemisphere, the Arctic is more exposed to atmospheric pollution of anthropogenic origin than is Antarctica, but we will see that this continent is not totally spared. Although heavy metals such as lead, coming from mining activity or from our fuel, have above all marked the Arctic, the greenhouse gases and the ozone hole have reached Antarctica. The composition of the atmosphere has been studied in a continuous fashion for only a few decades—since the International Geophysical Year of 1957–58 for carbon dioxide and more recently for other greenhouse gases or pollutants. The archives contained in the ice are unique and thus indispensable for putting the influence of human beings on their environment into perspective.

It is not always simple to pass from concentrations measured in the snow to those that exist in the atmosphere; however, researchers have been able to decipher the memory of the ice to account for the principal changes that have occurred throughout the years (since the beginning of the industrial age and sometimes beyond) from pits or shallow ice cores. Before going to Antarctica, let's take a final look at Greenland where researchers have reconstructed many scenarios of the state of the atmosphere.

The Story of Lead

The well-documented history of lead illustrates many facets of pollution. It was Clair Patterson of Caltech in California who was the first, in 1969, to use the archives of inland ice sheets to reconstruct the evolution of that pollutant.[1] With his colleague Murozumi, he discovered an increase in concentrations of lead in the snow of Greenland that was particularly marked at the beginning of the industrial era. The extraction process required extremely clean techniques; researchers had to pass into "white rooms" before using specific ultrasensitive methods of analysis. So, in snow from the last few thousand years, 1,000 tons of ice contained only 0.1 milligrams of lead, which came from the dust of rocks or from volcanoes, whereas there was close to 200 times more for the highest recent layers. Pollution seems in this case an inappropriate term, but it is the sensitivity of detection and the global tendency that interest us here. Claude Boutron and his French and international colleagues then improved the technological developments that would advance our knowledge in this domain.[2]

Thus the ice of Greenland bore witness to very old traces of lead in the atmosphere of the Northern Hemisphere at the height of the Roman Empire. It was, for example, quite marked in the samples dating back 2,000 years that were connected with the exploitation of mineral resources. We then observe a decrease in concentrations that could correspond to the fall of the Roman Empire prior to the year 1000, before increasing again in the Middle Ages. As historians have shown, these evolutions are linked to the quantities of ore extracted from mines. By measuring the different isotopes that make up lead, researchers have been able to identify the major sites of production such as that of Rio Tinto in Spain. Although all of these impurities were emitted on the ground in the form of dust, traces of them have been found thousands of kilometers away in the snow of Greenland; this was no doubt the first revelation of a trace of pollution extending over a large, hemispheric scale.

But the story of lead took another turn during the last two centuries. Although not very detailed, the curve of Patterson and his colleagues, published in the 1960s, demonstrates a strong increase in the amount of lead in the snow of northwestern Greenland, with levels as much as two hundred times higher than natural concentrations. These results had enormous repercussions; they demonstrated that humans were responsible for the undeniable pollution of the atmosphere of the Northern Hemisphere and were the basis for a crusade undertaken by researchers against the addition of lead in gas to improve the performance of automobile fuel. The responsibility of that additive for an increase in concentrations observed was confirmed by the measurement of "organoleads" in the snow that do not exist in the natural environment.

The use of leaded gas began in the 1930s and increased rapidly until around 1970 (Figure 17.1). Other measurements bearing on the isotopic composition of lead enabled us to identify the guilty parties of this pollution: American cars from the 1970s are at the origin of two-thirds of the lead present in the Greenland snow, whereas Europe was the essential source in the 1980s. The United States was the first to respond to the warning of the ice. As a result of pressure from ecologists, unleaded gas gradually caught on and we have returned to more natural amounts in the snow and air of Greenland.

In the great South, the story of lead recorded in the snow of Antarctica has a completely different slant. The concentrations are much weaker than those of Greenland and thus even more difficult to measure. It was only recently that the first traces of pollution from the end of the nineteenth century were identified in the coastal ice cores. These resulted from whaling activity, from ship traffic using coal, and from mining activity in the various continents of the South—far away from the inland ice sheet. But there was nothing alarming in that data.

Other Heavy Metals, Including Copper

From extractions taken in the center of Greenland at Summit Station, French scientists[3] were able to reconstruct the history of pollution of the Northern Hemisphere between 1773 and 1992 by analyzing so-called heavy metals. During the last two centuries, concentrations multiplied eight times for cadmium, five for zinc, and four for copper until the 1970s; these coefficients were reduced by about a factor of two during the following twenty years. Despite this improvement, biogeochemical atmospheric cycles were still largely dominated by emissions of human origin at thousands of kilometers from the chemical and mining industrial centers and from sources using fossil fuels such as coal.

The history of atmospheric pollution from copper revealed in the ice of Greenland has also been edifying.[4] It shows that concentrations began to exceed natural levels around 4,500 years ago with the rise in the metallurgy industry in the Copper and then the Bronze Age, which combined copper with tin. During the Greco-Roman era, around the fourth century B.C.E., concentrations were on average double natural levels; they remained

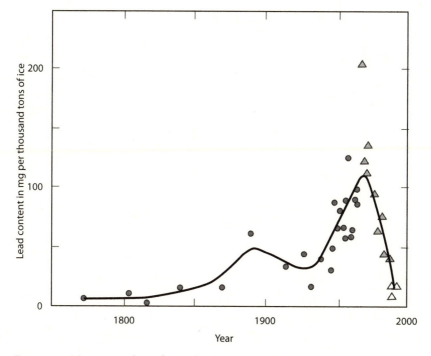

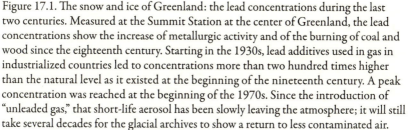

Figure 17.1. The snow and ice of Greenland: the lead concentrations during the last two centuries. Measured at the Summit Station at the center of Greenland, the lead concentrations show the increase of metallurgic activity and of the burning of coal and wood since the eighteenth century. Starting in the 1930s, lead additives used in gas in industrialized countries led to concentrations more than two hundred times higher than the natural level as it existed at the beginning of the nineteenth century. A peak concentration was reached at the beginning of the 1970s. Since the introduction of "unleaded gas," that short-life aerosol has been slowly leaving the atmosphere; it will still take several decades for the glacial archives to show a return to less contaminated air.

at that level during the Middle Ages before quickly rising after the industrial revolution. To explain these variations, researchers compiled data published on the production of copper throughout history. According to those data, production began in the Neolithic, around 7,000 years ago, but became significant only 2,000 years later. It then continued developing, culminating first at the height of the Roman Empire. Although copper production decreased in Europe during the medieval era, most of it came from China, notably during the Sung dynasty of the North (tenth to twelfth centuries A.D.)

during which a second maximum production close to that of the Roman era was reached. Production then decreased again before undergoing yet another strong growth into the present, where it is around nine million tons per year.

This simplified history coincides with the data from the ice, but beyond the raw figures of mining production, the glacial records also include the effect of the techniques used. Until the end of the eighteenth century, those were extremely polluting; 15% of copper emissions into the atmosphere were caused by production methods whereas currently they do not go beyond 0.25%. The data from the glacial archives open a possible (although difficult) path toward a quantitative approach to the history of the production of metals in ancient civilizations; this is an important socioeconomic and political parameter since it controlled in particular the minting of coins and the manufacture of weapons. Returning to recent pollution, the fallout measured in Greenland is on the same order as that in the twentieth and the nineteenth centuries even though mining activity has increased greatly; we see here the beneficial role of technological progress.

Sulfates

The level of atmospheric pollution in the Arctic is sometimes so high that "Arctic fog" is a problem. The term was first used in the 1950s to describe an unusual reduction in visibility that U.S. pilots had observed while flying over the high Arctic latitudes. That fog is seasonal, with a peak in the spring, and comes from anthropogenic sources of emissions located outside the Arctic. The fog aerosols are essentially sulfurated and come from the burning of coal in mid-latitudes.[5]

The sulfates present in the atmosphere come from both marine aerosols and burning coal and gas. The natural portion can be evaluated by measuring the elements characteristic of an ocean source, such as sodium. That enables the isolation of the component linked to pollution by subtracting the fallout from volcanic eruptions rich in sulfates that have been localized in time and are easily detectable.

At Summit Station (see Figure 17.2) concentrations increased at the beginning of the twentieth century until they peaked in the 1930s; that increase

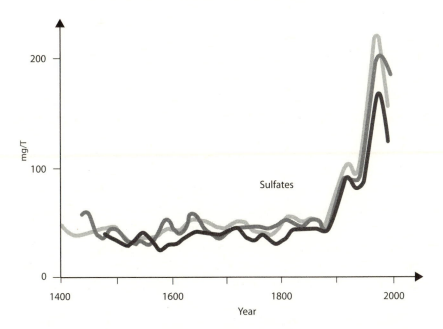

Figure 17.2. The snow and ice of Greenland: quantity of sulfates during the last centuries. The amounts measured in the snow and ice at Summit Station and in two other sites in Greenland show a remarkable similarity. The natural level, which was stable over the centuries, was greatly perturbed in the last years of the nineteenth century. The impact of human activity culminated before the end of the twentieth century; the decrease subsequently observed was not due to a decrease in industrial activity but to purification through filtration of emissions in our large cities.

was followed by a slight decrease, no doubt linked to the economic crisis. After World War II a much more rapid increase ensued, but filtration measures for emissions reflected in amounts measured after 1980 indicated a partial return to natural levels.

As does lead, sulfates of anthropogenic origin behave differently in Antarctica and Greenland. They increased in Greenland from a factor of three to a factor of five in two stages centered on the 1900s and the 1950s. By contrast, in Antarctica, the quantities of sulfates have varied little over the past hundred years, which probably indicates the distance of the sources and the short time (only a few days) that sulfates stay in the atmosphere.

Radioactivity

A striking example of how our atmosphere is connected on a global scale is that of radioactive dust. In the two ice sheets of Antarctica and Greenland, the radioactivity of snow layers has faithfully recorded the calendar and the power of nuclear explosions in the atmosphere. Just as in the case of strong volcanic eruptions, the energy freed during nuclear explosions is sufficient for the radioactive elements to reach the stratosphere and, from there, fall over the entire globe. The profile obtained at the South Pole at the end of the 1970s is typical of the variations observed over the entire continent and is for us a revelation showing the fact that for some pollutants we have only one atmosphere on a global scale.[6] The natural beta radioactivity of the layers of snow was at the time around 70 disintegrations per hour and per kilogram of snow (dhp/h); it came principally from lead 210 and from its descendants. Compared to that basic natural noise, the levels of radioactivity are most often much higher. In the spring of 1955 a peak was observed that represented the fallout from the first thermonuclear tests in March 1954. Radioactivity increased by a factor of twenty and remained great until around 1965, during which the fallout from the series of large explosions carried out in the Northern Hemisphere between September 1961 and December 1962 (close to three years earlier) were deposited.

Certain layers of snow containing the fallout of large nuclear explosions were showing radioactive levels up to forty times greater than the natural (in the absence of fallout from nuclear explosion) radioactive level. The suspension of nuclear testing in the atmosphere in 1963–66 was seen in a decrease in radioactivity connected to the depletion of the stratospheric reservoir. Strontium 90 and cesium 137 were the primary causes of beta radioactivity. These are elements whose half-lives are close to twenty years and, with the passing of time, the measurements will be much less impressive.

A series of secondary peaks was observed in the snow deposited between 1969 and 1976; they correspond to a partial resumption of atmospheric tests, whereas the radioactivity returned to values close to the natural level. The particularly high 1955 and 1965 levels enabled the description, over all of Antarctica, of the layers deposited during those years and serve as markers to measure the accumulation of snow. Similarly, the fallout of tritium is also recorded in Antarctic ice.[7] A radio-element produced naturally by the action of cosmic radiation over compounds in the atmosphere, it is also formed during thermonuclear explosions; it was a hundred times more concentrated in the snow deposited around the end of 1965 than at the beginning of the 1950s.

The glacial archives thus measure the amount of pollution in our atmosphere far from human sources of emission. They give us a point of reference for the natural "background noise" from which we can evaluate human impact. As for aerosols, the response is clear: humans have marked the atmosphere, from the North Pole to the South Pole. But what about gases?

The Ozone Hole: An Emblematic Pollution

The ozone molecule, formed of three oxygen atoms, is not very abundant in the atmosphere. In the upper atmosphere (stratosphere), where its concentration is higher, only three or four molecules in a million are ozone molecules. Closer to home, in the lower atmosphere (troposphere), the proportion of this component is no more than a few dozen billionths and represents only around 10% of the total mass of ozone in the atmosphere. The ozone of the lower atmosphere is, however, a key component of atmospheric chemistry, ruling in particular the oxidizing capacity of the atmosphere and thus the duration of life and therefore the abundance of many other greenhouse gases, such as methane.

Absorbing infrared radiation emitted by the Earth, ozone is directly responsible, as we have seen, for an additional greenhouse effect of 0.35 Wm^{-2}, or 20% to 25% of the contribution of carbon dioxide. It also controls the thermal structure of the stratosphere within which an absorption of ultraviolet radiation by the ozone constitutes the principal source of warming. Various pollutants can contribute indirectly to the climatic perturbations through photochemical processes that cause, for example, the production of ozone in the troposphere to intervene. This is the case of carbon monoxide (CO) and nitrogen oxides (NO_X), 80–90% of which are emitted through the use of fossil fuels (industry, transportation) or by burning the biomass (deforestation, heating wood, bush fires). The emission of these components from the ground or their injection directly into the atmosphere, for example, in the case of NO_X emitted by airplanes, leads to the photochemical production of O_3. This pollution is limited to the troposphere and is redistributed by winds, increasing the basic level in the atmosphere far from the zones of emission. The ozone molecule has a relatively short life in the troposphere, on the order of a few weeks on average, and the distribution of ozone is thus quite variable in space and in time, with a strong seasonal variation and a heterogeneous geographic distribution.

The concentration of ozone on the surface has been followed at many sites for several dozen years, but measurements before the 1970s are rare. Those done at the observatory of Parc Montsouris in Paris (1876–1909) and Pic du Midi in the Pyrénées (1874–1909) indicate an order of grandeur of ten-billionths, but we had to wait for the measurements taken in the Northern Hemisphere in the 1950s to clearly show an increase in the existing amount. It is now estimated that it has more than tripled during the twentieth century. A maximum ozone reaching more than sixty-billionths in the summer is calculated by the chemical models above polluted regions of the Northern Hemisphere (North America, Europe, Southeast Asia). These models estimate that the global increase in tropospheric ozone has been on the order of 30% since the beginning of the preindustrial era. It is a pity that the ice, which does not retain the memory of these variations, does not allow us to validate that estimate.

Although the ozone in the air that we breathe on the ground is toxic to our health, that which is more abundant in the stratosphere is more vital. Because it absorbs the solar ultraviolet radiation of short wavelength before it reaches the Earth, ozone protects life on our planet. Amassed together and concentrated in average conditions on the Earth's surface, it would have a thickness of only three millimeters, or 300 Dobson units in scientific jargon. It is surprising that such a thin layer protects life from ultraviolet radiation that can cause mutations of DNA and skin cancer; without ozone, life would not be possible except in the ocean depths.

The ozone hole in the stratosphere was discovered in Antarctica in 1985.[8] The CFCs emitted at the time by the developed and industrialized countries of the Northern Hemisphere were responsible for this, demonstrating the sometimes unpredictable impact of human activity on the entire environment in which we live.

The seasonal depletion of stratospheric ozone first observed above Antarctica at the end of the polar night and more recently above the Arctic is one of the principal problems affecting our environment. The thickness, the surface area, and the duration of the Antarctic ozone hole has continuously increased, reaching an apogee of nearly 30 million km^2 in the early 2000s with ozone concentrations about two times less than what they were in the 1960s.

In the Arctic, the ozone hole is less noticeable following a more moderate cooling of the stratosphere, but the mean annual levels of ozone decreased by 10% in the 1990s compared to levels at the end of the 1970s, which increases the risk to populations who live there.

The Montreal Protocol signed in 1987 addressed the protection of the ozone layer; it was a matter of halting its spread and reducing its intensity. This treaty, refined on several occasions, aimed primarily to drastically reduce CFC emissions; when 190 countries gathered again in Montreal in 2007, they noted the effectiveness of the measures proposed. This does not mean the battle has been won; it will certainly require several decades for our atmosphere to regain its natural state, which has been particularly degraded in the polar regions, but we can rejoice at this positive sign for our environment following a concerted action on an international level.

The Anthropocene and Greenhouse Gases

Unlike aerosols, whose concentrations decrease the farther they are from their source, unreactive gases have the same concentration everywhere in our atmosphere, and the industrial period, as we have seen, is strongly marked by a rapid increase in the amount of greenhouse gases. The analysis of air bubbles contained in the ice shows that since 1750 concentrations of CO_2 have increased by almost 40% and those of CH_4 have more than doubled. These bubbles also teach us that the current amounts are much higher than those of the last hundreds of thousands of years. The industrial period marks the beginning of an era characterized by the impact of humans on the natural environment. It was the increase in concentrations of carbon dioxide and methane starting at the end of the eighteenth century, as is recorded in polar ice, which enabled Paul Crutzen,[9] Nobel Prize winner in physics in 1995, to define the beginning of the Anthropocene Era.

The Anthropocene Era

Climate warming has become one of the major challenges that our global society must face. The discoveries made in the ice sheets, notably in Antarctica, have proven this to scientists worldwide. In addition, field data and satellite observations, as well as the viability of the models used for analyzing climate change, have shown us that we cannot ignore this phenomenon. For our part, we have wanted to concentrate on our white planet, whose role in this challenge is crucial since the polar ice is both a unique witness and an essential actor.

Homo habilis, which appeared 2.5 million years ago, then *Homo erectus* and Cro-Magnon Man had to defend themselves against nature to survive. In that long struggle the paths taken in the migrations of our ancestors were the precursors of our highways. The first human beings fed themselves by gathering and hunting, and then, after becoming more sedentary, they cultivated the land and developed an agriculture adapted to their needs. Animal farming succeeded hunting, and to feed those animals they had to cultivate artificial prairies. We thus entered into the Holocene, a new era for the human way of life but also for the climate which, around 10,000 years ago, was already in a warm condition. To warm themselves and cook food, humans burned wood from forests; with the great leap of the conquest of fire began the still undetectable production of greenhouse gas emissions of anthropogenic origin.

Gradually the landscapes changed. Thus through the millennia humans built structures and cities, visibly reducing the extent of natural spaces. In doing so they disturbed natural geochemical cycles and, beyond the land, our atmosphere was also affected. From wood we moved to fossil fuel, coal, oil, and gas to satisfy the energy demands of a population that was growing rapidly at the same time that the industrial era began. We then entered the Anthropocene, characterized by a rise in pollution. Humans thus strongly marked the atmosphere and oceans of the Earth. Humans have become the greatest predator of our environment and a major actor in its destruction, so great is their influence on the evolution of our climate, biodiversity, and, more generally, our living conditions. This destruction cannot be masked by technological and medical progress, which, for example, enables people to live longer.

Among the signs of this destruction, the warning from polar regions, which are practically uninhabited and far from the sources of pollution, is particularly impressive. Two striking examples are the ozone hole, which was discovered by probing the upper atmosphere above Antarctica, and the sudden increase in greenhouse gases. These two symbolic examples plead for a balanced support of research. Although it is necessary to finance projects that demonstrate precise objectives in an aim toward solving social problems, we also need to strongly support basic and quality research already in progress and whose outcome is often unpredictable. This is what happened during the ozone discovery made by the English while they were taking routine measurements, a discovery that seemed unbelievable to the researchers; the first sets of data were even thrown away. This is also the case for research undertaken on glacial ice cores. Long years of observation during winters spent on a base, exploratory fieldwork and extractions, ice core campaigns and measurements in the laboratory by means of sophisticated techniques—all of these efforts have resulted in an awareness of the importance of that ice for the reconstruction of the climate and the composition of the atmosphere. Such research has opened a door onto a critical problem facing our society; we are proud to have helped open this door to the future.

Climate warming and its impacts have become serious realities that foretell of a very different world to which humans and their civilizations will have to adapt, not without damage. In the face of that world and, more generally, of environmental problems, we must clear the path for a more harmonious cohabitation between man and that environment, with a view toward safeguarding the future of generations to come, a notion that the expression "sustainable development" sums up in part.

We have only one planet. The atmosphere and the ocean have no other frontiers. The struggle against climate warming based on controlling greenhouse gas emissions demands a global solidarity involving all human beings, from states to citizens. That solidarity is difficult to achieve, given the differences among those involved, whose behavior too often favors the short term and responds only to their own interests, whether on the level of the individual or on that of industrial, financial, political, or even state groups. And yet it is time to act on an international level through novel initiatives such as those that have been put in place in France like the Grenelle de l'environnement.

Getting people to act will be difficult given budgetary constraints and the fact that right now people are being asked to drastically change their behavior based on events occurring far away. Contemplating the snow on the Belledonne chain or the midnight sun on the ice sheets does not convey the sense that the Earth is in danger or the urgency of the measures that need to be taken.

The control by humans, a living species among so many others, over the global environment now has a label: the Anthropocene. We have already used this term, which may not have been familiar to the reader, but it is good to bring it up again. The Anthropocene signifies the passage, in the evolution of the environment of our planet, from a balance that depended essentially on natural causes to a state in which the influence of humans becomes the governing factor.

If we go back to the beginning of the Quaternary, around 1.6 million years ago, warming and cooling depended on long periodicities of the Earth's trajectory around the Sun. During that time, the biosphere, that is, all ecosystems, in which humans then occupied very little space, participated in the saga of the climate. As we have seen, the variation in the composition of the atmosphere modified the radiative levels and through that even the climate. The gases that intervened in these levels depended in large part on geochemical cycles regulated by living species present in the ground, the oceans, and the atmosphere. Among those species, humans, through their activities, imposed their imprint, as is seen in detail in the evolution of the composition of the atmosphere as recorded in the ice.

These data led Paul Crutzen to write: "The Anthropocene could be said to have started in the late eighteenth century, when analyses of air trapped in polar ice showed the beginning of growing global concentrations of carbon dioxide and methane."[1] This statement clearly conveys the message from the ice.

Regarding the climate, the Anthropocene will no doubt mark our future, in any case for the century we are entering and likely the following ones. The climate and the environment are opening the path to many dangers, to many conflicts. In a near future, the rarity and the cost of energy resources will perhaps grant a certain respite to our environment, but they will plunge our world into the social problems born in the Anthropocene. If the Nobel Peace Prize was awarded to Al Gore and the IPCC, it was because beyond the quality of scientific expertise, climate warming creates a state of instability that can lead to migrations and conflicts provoked by unequal access to resources (water, food, energy), by sea-level rise, and by an increase in poverty, not to mention the difficulties linked to religions or to intransigent cultures. Will wars also mark the Anthropocene, or will we be able to find the complex recipes to fashion a peaceful planet?

We hope that the reader will have recognized the path illustrated by the following quotation from Théodore Monod, a great explorer of the hot deserts: "Until the nineteenth century, scientists were adventurers in the noble sense of the term, because the exploration of the planet hadn't ended. . . . . There are no more islands to discover; we must now seek to know how the world that surrounds us functions and above all how man will behave with regard to this little, so fragile ball turning in the immensity of the universe."[2]

Our behavior "with regard to this little ball" is truly the question raised at present by climate warming. A recent report of the UN on the environment (Nairobi, October 25, 2007)[3] made it a priority among the problems that our civilization must resolve. In this context, "the destiny of polar regions is crucial for the planet." We hope that readers have reached this conclusion from reading this book.

Let us repeat: climate warming is one of the great challenges facing our civilization today, and the polar ice is a witness to and an essential actor in it. These are good reasons for ice researchers to be concerned, well beyond the recent International Polar Year, with the state of health of the polar regions, those sentinels of our environment.

NOTES

Preface

1. Jared Diamond, *Collapse: How Societies Choose to Fail or Succeed* (New York: Viking, 2004), 212.

2. Ibid., 21.

Chapter 1

1. John Mercer, "West Antarctic Ice Sheet and CO_2 Greenhouse Effect: A Threat of Disaster," *Nature* 271 (1978): 321–25.

Chapter 2

1. Bernard Francou and Christian Vincent, *Les glaciers à l'épreuve du climat* (Belin : IRD Éditions, 2007).

2. Chester Langway, "The History of Early Polar Ice Cores," *Cold Regions Science and Technology* (2008), doi: 10016/j.coldegions.2008.01.001.

3. Michiel van den Broecke et al., "Partitioning Recent Greenland Mass Loss," *Science* 326 (2009): 984–86.

4. Isabella Velicogna et al., "Increasing Rates of Ice Mass Loss from the Greenland and Antarctic Ice Sheets Revealed by GRACE," *Geophysical Research Letters* 36 (2009): L19503.

Chapter 3

1. John Imbrie and Katherine Palmer Imbrie, *Ice Ages: Solving the Mystery* (Cambridge, MA: Harvard University Press, 1986).

2. Joseph Kirschvink, *The Proterozoic Biosphere: A Multidisciplinary Study*, ed. J. W. Schopf and C. C. Klein (Cambridge: Cambridge University Press, 1992), 51–52.

3. Yannick Donnadieu et al., "A 'Snowball Earth' Climate Triggered by Continental Break-up through Changes in Runoff," *Nature* 428 (2004): 303–6.

4. Robert de Conto and David Pollard, "A Coupled Climate-Ice Sheet Modeling Approach to the Early Cenozoic History of the Antarctic Ice Sheet," *Palaeogeography, Palaeoclimatology, Palaeoecology* 198 (2003): 39–52.

5. Milutin Milankovitch, *Kanon der Erdbestrahlung und Seine Andwendungauf das Eiszeitenproblem*, Royal Serbian Academy Special Publications 33 (Belgrade: Mihaila Čurčića, 1941): 132; translated into English in 1969.

6. Jim Hays, John Imbrie, and Nick Shackleton, "Variations in the Earth's Orbit: Pacemakers of the Ice Ages," *Science* 194 (1976): 1121–32.

Chapter 4

1. Claude Lorius and Liliane Merlivat, "Isotopes and Impurities in Snow and Ice," in *Proceedings of the Grenoble Symposium Aug.–Sept. 1975* (Vienna: IAHS, 1977), 125–37.

2. Jean Jouzel and Liliane Merlivat, "Deuterium and Oxygen 18 in Precipitation: Modeling of the Isotopic Effect during Snow Formation," *Journal of Geophysical Research* 89 (1984): 11749–57.

3. Gordon Manley, "Central England Temperatures: Monthly Means, 1659–1673," *Quarterly Journal of the Royal Meteorological Society* 101, no. 428 (1974): 389–405.

4. Jean-Marc Moisselin et al., "Changements climatiques en France au XXᵉ siècle: Étude des longues séries de données homogénéisées françaises de précipitations et températures," *La Météorologie* 38 (2002): 45–56.

5. Emmanuel Le Roy Ladurie, *Histoire humaine et comparée du climat*, vol. 1, *Canicules et glaciers* (Paris: Fayard, 2004) and vol. 2, *Disettes et révolutions* (Paris: Fayard, 2006).

6. Isabelle Chuine et al., "Grape Ripening as a Past Climate Indicator," *Nature* 432 (2004): 289–90.

7. Valérie Masson-Delmotte et al., "Changes in European Precipitation Seasonality and in Drought Frequencies Revealed by a Four-Century-Long Tree-Ring Isotopic Record from Brittany, Western France," *Climate Dynamics* (2005), doi: 10.1007/s00382-004-0458-1.

8. Joël Guiot et al., "A 140,000-Year Climatic Reconstruction from Two European Pollen Records," *Nature* 338 (1989): 309–13.

9. Jacques-Louis de Beaulieu and Maurice Reille, "A Long Upper Pleistocene Pollen Record from Les Echets, near Lyon, France," *Boreas* 13 (1984): 111–32.

10. Maurice Reille and Jacques-Louis de Beaulieu, "Long Pleistocene Pollen Record from the Praclaux Crater, South-Central France," *Quaternary Research* 44 (1995): 205–15.

11. Denis Didier-Rousseau, "Paleoclimatology of the Achenheim Series: A Malacological Analysis," *Palaeogeography, Palaeoclimatology, Palaeocology* 59 (1987): 293–314.

12. Uli Von Grafenstein et al., "A Mid-European Decadal Climate Record from 15,500 to 5,000 Years BP," *Science* 284 (1999): 1654–57.

13. Yongjin Wang et al., "Millennial and Orbital-Scale Changes in the East Asian Monsoon over the Past 224,000 Years," *Nature* 451 (2008): 1090–93.

14. Dominique Genty et al., "Precise Dating of Dansgaard-Oeschger Climate Oscillations in Western Europe from Speleothem Data," *Nature* 421 (2003): 833–37.

15. Édouard Bard et al., "Calibration of the ^{14}C Time Scale over the Past 30,000 Years Using Mass Spectrometic U/Th Ages from Barbados Corals," *Nature* 345 (1990): 405–10.

16. Nick Shackleton, André Berger, and Dick Peltier, "An Alternative Astronomical Calibration of the Lower Pleistocene Timescale Based on ODP Site 677," *Transactions of the Royal Society of Edinburgh* 81 (1990): 251–70.

17. Jurg Luterbacher et al., "European Seasonal and Annual Temperature Variability," *Science* 303 (2004): 1499–1503.

Chapter 5

1. Philippe Ciais and Jean Jouzel, "Deuterium and Oxygen 18 in Precipitation: An Isotopic Model Including Mixed Cloud Processes," *Journal of Geophysical Research* 99 (1994): 16793–803.

2. Jean Jouzel et al., "Water Isotopes in Precipitation: Data/Model Comparison for Present-Day and Past Climates," *Quaternary Science Reviews* 19, no. 1-5 (2000): 363–79.

3. Françoise Vimeux et al., "New Insights into Southern Hemisphere Temperature Changes from Vostok Ice Cores Using Deuterium Excess Correction over the Last 420,000 Years," *Earth and Planetary Science Letters* 203 (2002): 829–43.

4. Jeff Severinghaus et al., "Timing of Abrupt Climate Change at the End of the Younger Dryas Interval from Fractionated Gases in Polar Ice," *Nature* 391 (1998): 141–46.

5. Michael Bender et al., "The Dole Effect and Its Variation during the Last 130,000 Years as Measured in the Vostok Core," *Global Biogeochemical Cycle* 8 (1994): 363–76.

6. Dominique Raynaud et al., "The Local Insolation Signature of Air Content in Antarctic Ice: A New Step toward an Absolute Dating of Ice Records," *Earth and Planetary Science Letters* 261 (2007): 337–49.

7. Grant Raisbeck et al., "Evidence for Two Intervals of Enhanced Deposition in Antarctic Ice during the Last Glacial Period," *Nature* 326 (1987): 273–77.

8. Grant Raisbeck et al., "Absolute Dating of the Last 7,000 Years of the Vostok Ice Core Using ¹⁰Be," *Mineralogical Magazine* 62A (1998): 1228.

9. Jean Jouzel et al., "Climatic Interpretation of the Recently Extended Vostok Ice Records," *Climate Dynamics* 12 (1996): 513–21.

10. Frédéric Parrenin et al., "Dating the Vostok Ice Core by an Inverse Method," *Journal of Geophysicsal Research* 106 (2001): 31837–51.

11. Michael Bender, "Orbital Tuning Chronology for the Vostok Climate Record Supported by Trapped Gas Composition," *Earth and Planetary Science Letters* 204 (2002): 275–89.

Chapter 6

1. Willi Dansgaard, "Stable Isotopes in Precipitation," *Tellus* 16 (1964): 436–68.

2. Willi Dansgaard et al., "Isotopic Distribution in a Greenland Iceberg," *Nature* 185 (1960): 232.

3. Willi Dansgaard et al., "One Thousand Centuries of Climatic Record from Camp Century on the Greenland Ice Sheet," *Science* 166 (1969): 377–81.

4. Sigfus J. Johnsen, Hank B. Clausen, Willi Dansgaar, and Chester Langway, "Oxygen Isotope Profiles through Antarctic and Greenland Ice Sheets," *Nature* 235, no. 5339 (1972): 429–34, doi: 10.1038/235429a0.

5. Jean Jouzel, Liliane Merlivat, Michel Pourchet, and Claude Lorius, "A Continuous Record of Artificial Tritium Fallout at the South Pole (1954–1978)," *Earth and Planetary Science Letters* 45 (1979): 188–200.

6. E. Picciotto, X. De Maere, and I. Friedman, "Isotope Composition and Temperature of Formation of Antarctic Snows," *Nature* 187 (1960): 857–59.

7. Claude Lorius, *Les glaces de l'Antarctique* (Paris: Odile Jacob, 1991).

8. Claude Lorius and Liliane Merlivat, "Distribution of Mean Surface Stable Isotope Values in East Antarctica: Observed Changes with Depth in a Coastal Area, in Isotopes and Impurities in Snow and Ice," *Proceedings of the Grenoble Symposium* 118 (1977): 127–37.

9. Dominique Raynaud and Claude Lorius, "Climatic Implications of Total Gas Content in Ice at Camp Century," *Nature* 243 (1973): 283–84.

10. Jean Jouzel, Liliane Merlivat, and Etienne Roth, "Isotopic Study of Hail," *Journal of Geophysical Research* 80 (1975): 5015–30.

11. Claude Lorius et al., "A 30,000 Year Isotope Climatic Record from Antarctic Ice," *Nature* 280 (1979): 644–48.

12. Robert Delmas et al., "Polar Ice Evidence That Atmospheric CO_2 20,000 BP Was 50% of Present," *Nature* 284 (1980): 155–57; Albrecht Neftel et al., "Ice Core Sample Measurements Give Atmospheric CO_2 Content during Past 40,000 Years," *Nature* 295 (1982): 220–23.

13. Jean-Robert Petit et al., "Ice Age Aerosol Content from East Antarctic Ice Core Samples and Past Wind Strength," *Nature* 293 (1981): 391–94.

14. Paul Duval and Claude Lorius, "Crystal Size and Climatic Record Down to the Last Ice-Age from Antarctic Ice," *Earth and Planetary Science Letters* 48 (1980): 59–64.

15. Michel Legrand et al., "Vostok (Antarctica) Ice Core, Atmospheric Chemistry Change over the Last Climatic Cycle (160,000 Years)," *Atmospheric Environment* 22 (1988): 317–31.

16. Jean Jouzel, Liliane Merlivat, and Claude Lorius, "Deuterium Excess in an East Antarctic Ice Core Suggests Higher Relative Humidity at the Oceanic Surfaces during the Last Glacial Maximum," *Nature* 299, no. 5885 (1982): 588–91.

17. Françoise Yiou et al., "Be-10 in Ice at Vostok Antarctica during the Last Glacial Cycle," *Nature* 316 (1985): 616–17.

18. Willi Dansgaard et al., "A New Greenland Deep Ice Core," *Science* 218 (1982): 1273–77.

19. Claude Lorius et al., "A 150,000-Year Climatic Record from Antarctic Ice," *Nature* 316 (1985): 591–96.

20. Jean Jouzel et al., "Vostok Ice Core: A Continuous Isotope Temperature Record over the Last Climatic Cycle (160,000 Years)," *Nature* 329 (1987): 402–8; Jean-Marc Barnola, "Vostok Ice Core Provides 160,000-Year Record of Atmospheric CO_2," *Nature* 329 (1987): 408–14; Christophe Genthon et al., "Vostok Ice Core: Climatic Response to CO_2 and Orbital Forcing Changes over the Last Climatic Cycle," *Nature* 329 (1987): 414–18.

21. We thank Wally Broecker for sending Jean Jouzel the report of the meeting in Boston.

22. Hideaki Motoyama, "The Second Deep Ice Coring Project at Dome Fuji, Antarctica," *Scientific Drilling* 5 (2007): 41–43.

23. Barbara Stenni et al., "Unified Antarctic and Greenland Climate Seesaw during the Last Deglaciation," *Nature Geosciences* 4 (2011): 46–49.

24. Eric Steig et al., "A High Resolution Stable Isotope Record from Central West Antarctica Covering the Last 62,000 Years," *Geophysical Research Abstracts* 14, EGU2012-10419-1 (2012).

25. North Greenland Ice Core Project Members, "High Resolution Record of Northern Hemisphere Climate Extending into Last Interglacial Period," *Nature* 431 (2004): 147–51.

26. Dorthe Dahl-Jensen and the NEEM Community, "The NEEM Climate Record" (paper presented at the European Geosciences Union annual meeting, Vienna, April 23–27, 2012).

27. Jean Jouzel et al., "Orbital and Millennial Antarctic Climate Variability over the Past 800,000 Years," *Science* 317 (2007): 793–96.

Chapter 7

1. CLIMAP, "The Surface of the Ice-Age Earth," *Science* 191 (1976): 1131–37.

2. Michael Bender et al., "Isotopic Composition of Atmospheric O_2 in Ice Linked with Deglaciation and Global Primary Productivity," *Nature* 318 (1985): 349–52.

3. Jean Jouzel et al., "Deuterium Excess in an East Antarctic Ice Core Suggests Higher Relative Humidity at the Oceanic Surface during the Last Glacial Maximum," *Nature* 299 (1982): 588–91.

4. Delmas et al., "Polar Ice Evidence That Atmospheric CO_2 20,000 BP Was 50% of Present," 155–57.

5. Jérôme Chappellaz et al., "Ice-Core Record of Atmospheric Methane over the Past 160,000 Years," *Nature* 345 (1990): 127–31.

6. Claude Lorius et al., "Greenhouse Warming, Climate Sensitivity and Ice Core Data," *Nature* 347 (1990): 139–45.

7. Jean-Robert Petit et al., "Climate and Atmospheric History of the Past 420,000 Years from the Vostok Ice Core, Antarctica," *Nature* 399 (1999): 429–36.

8. Jean-Robert Petit et al., "Paleoclimatological Implications of the Vostok Core Dust Record," *Nature* 343 (1990): 56–58.

9. Francis Grousset et al., "Antarctic Ice Core Dusts at 18 Kyr BP: Isotopic Constraints on Origin and Atmospheric Circulation," *Earth and Planetary Science Letters* 111 (1992): 175–82.

10. Dominique Raynaud et al., "The Record for Marine Stage 11," *Nature* 4367 (2005): 39–40.

11. Jean Jouzel et al., "More than 200m Thick of Lake Ice above the Subglacial Lake Vostok, Antarctica," *Science* (1999): 2138–41.

Chapter 8

1. Okitsugu Watanabe et al., "Homogeneous Climate Variability across East Antarctica over the Past Three Glacial Cycles," *Nature* 422 (2003): 509–12.

2. Gabrielle Dreyfus et al., "Anomalous Flow below 2,700m in the EPICA Dome C Ice Core Detected Using d18O of Atmospheric Oxygen Measurements," *Climate of the Past* 3 (2007): 341–53.

3. Frédéric Parrenin et al., "The EDC3 Chronology of the EPICA Dome C Ice Core," *Climate of the Past* 3 (2007): 485–97.

4. Grant Raisbeck et al., "Be-10 Evidence for the Matuyama-Brunhes Geomagnetic Reversal in the Dome C Ice Core," *Nature* 444 (2006): 82–84.

Chapter 9

1. Willi Dansgaard, Jim White, and Sigfus Johnsen, "The Abrupt Termination of the Younger Dryas Climate Event," *Nature* 339 (1989): 532–34.

2. Wally Broecker et al., "Does the Ocean-Atmosphere System Have More than One Mode of Operation?" *Nature* 315 (1985): 21–26.

3. Hans Oeschger et al., "Late Glacial Climate History from Ice Cores, in Climate Processes and Climate Sensitivity," in J. E. Hansen and T. Takahashi, eds., *American Geophysical Union* (Washington, DC: NASA, 1984), 299–306.

4. Sigfus Johnsen et al., "Irregular Glacial Interstadials Recorded in a New Greenland Ice Core," *Nature* 359 (1992): 311–13.

5. Édouard Bard et al., "Calibration of the C-14 Timescale over the Past 30,000 Years Using Mass-spectrometric U-Th Ages from Barbados Corals," *Nature* 345 (1990): 405–10.

6. Richard Alley et al., "Abrupt Increase in Greenland Snow Accumulation at the End of the Younger Dryas Event," *Nature* 362 (1993): 527–29.

7. Uli von Grafenstein et al., "A Mid-European Decadal Isotope-Climate Record from 15,500 to 5,000 Years B.P.," *Science* 284 (1999): 1654–57.

8. North GRIP Ice Core Project Members, "High Resolution Record of Northern Hemisphere Climate Extending into the Last Interglacial Period," *Nature* 431 (2004): 147–51.

9. Willi Dansgaard et al., "Evidence for General Instability of Past Climate from a 250-kyr Ice-Core Record," *Nature* 364 (1993): 218–20.

10. GRIP Project Members, "Climatic Instability during the Last Interglacial Period Revealed in the Greenland Summit Ice-Core," *Nature* 364 (1993): 203–7.

11. Pieter Grootes et al., "Comparison of the Oxygen Isotope Records from the GISP2 and GRIP Greenland Ice Cores," *Nature* 366 (1993): 552–54.

12. Kurt Cuffey et al., "Large Arctic Temperature Change at the Wisconsin-Holocene Glacial Transition," *Science* 270 (1995): 455–58; Sigfus Johnsen, S.J., "Greenland Paleotemperatures Derived from GRIP Bore Hole Temperature and Ice Core Isotope Profiles," *Tellus* 47B (1995): 624–29.

13. Jeff Severinghaus et al., "Timing of Abrupt Climate Change at the End of the Younger Dryas Interval from Thermally Fractionated Gases in Polar Ice," *Nature* 391 (1998): 141–46.

14. C. Lang et al., "16°C Rapid Temperature Variation in Central Greenland, 70,000 Years Ago," *Science* 286 (1999): 934–37.

15. Amaelle Landais et al., "Large Temperature Variations over Rapid Climatic Events in Greenland: A Method Based on Air Isotopic Measurements," *CRAS* 377 (2005): 947–56.

16. Gerry Bond et al., "Correlations between Climate Records from North Atlantic Sediments and Greenland Ice," *Nature* 365 (1993): 143–47.

17. Valérie Masson-Delmotte et al., "GRIP Deuterium Excess Reveals Rapid and Orbital Changes of Greenland Moisture Origin," *Science* 309 (2005): 118–21.

18. Jorgen Peder Steffensen et al., "High-Resolution Greenland Ice Core Data Show Abrupt Climatic Change Happens in a Few Years," *Science* 321 (2008): 684.

19. EPICA Community Members, "One-to-One Coupling of Glacial Climate Variability in Greenland and Antarctica," *Nature* 444 (2006): 195–98.

20. Thomas Stocker and Sigfus Johnsen, "A Minimum Thermodynamic Model for the Bipolar Seesaw," *Paleoceanography* 18 (2003): 1087.

Chapter 10

1. Uli von Grafenstein et al., "The Short Period 8,200 Years Ago Documented in Oxygen Isotope Records of Precipitation in Europe and Greenland," *Climate Dynamics* 14 (1998): 73–81.

2. Édouard Bard et al., "Solar Irradiance during the Last Millennium Based on Cosmogenic Nucleides," *Tellus* 52B (2000): 985–92.

3. Bill Ruddiman, "The Anthropogenic Greenhouse Era Began Thousands of Years Ago," *Climatic Change* 61 (2003): 261–93.

Chapter 11

1. Charles D. Keeling, "Rewards and Penalties of Monitoring the Earth," *Annual Review of Energy and the Environment* 23 (1998): 25–82.

2. IPCC, *Climate Change 2007: The Physical Science Basis*, Working Group I Contribution to the Fourth Assessment Report of the IPCC (Cambridge: Cambridge University Press, 2007).

3. http://www.globalcarbonproject.org/.

4. http://www.globalcarbonproject.org/carbonbudget/index.htm.

Chapter 12

1. Willi Dansgaard et al., "Climate Changes, Norsemen, and Modern Man," *Nature* 255 (1975): 24–28.

2. John Mitchell, "The Greenhouse Effect and Climate Change," *Review of Geophysics* 27 (1989): 115–39.

3. J. D. Houghton et al., eds., *Climate Change 1995, The Science of Climate Change* (Cambridge: Cambridge University Press. 1995).

4. Michael Mann et al., "Northern Hemisphere Temperatures during the Last Millennium: References, Uncertainties, and Limitations," *Geophysical Research Letters* 26 (1999): 759–62.

5. IPCC, *Climate Change 2001: The Scientific Basis*, Third Assessment Report, summary for policymakers, 2001.

6. S. Solomon et al., eds., *Contribution of Working Group I to the Fourth Assessment Report of the Intergovernmental Panel on Climate Change* (Cambridge: Cambridge University Press, 2007).

7. Ibid.

8. Claude Allègre (with Dominique de Montvalon), *L'imposture climatique ou la fausse écologie* (Paris: Plon, 2010).

9. Vincent Courtillot, *Nouveau voyage au centre de la terre* (Paris: Odile Jacob, 2009).

10. Sylvestre Huet, *L'imposteur c'est lui: Réponse à Claude Allègre* (Paris: Stock, 2010).

11. "Climate Change and the Integrity of Science," *Science* 328 (2010): 689–90.

12. N. Caillon et al., "Timing of Atmospheric CO_2 and Antarctic Temperature Changes across Termination-III," *Science* 299 (2003): 1728–31.

13. Ibid.

14. Rune Graversen et al., "Vertical Structure of Recent Arctic Warming," *Nature* 541 (2008): 53–56.

15. Éric Rignot and Pannir Kanagaratnam, "Changes in the Velocity Structure of the Greenland Ice Sheet," *Science* 311 (2006): 986–90.

16. Göran Ekström et al., "Seasonality and Increasing Frequency of Greenland Glacial Earthquakes," *Science* 311 (2006): 1756–58.

17. Eugene Domack et al., "Stability of the Larsen B Ice Shelf on the Antarctic Peninsula during the Holocene Epoch," *Nature* 436 (2005): 681–85.

18. Éric Rignot et al., "Recent Antarctic Ice Mass Loss from Radar Interferometry and Regional Climate Modeling," *Nature Geoscience* 1 (2008): 106–10.

Chapter 13

1. Richard Lindzen et al., "Does the Earth Have an Adaptive Iris?" *Bulletin of the American Meteorological Society* 82, no. 3 (2001): 417–32.

2. Stefan Rahmstorf et al., "A Semi-Empirical Approach to Projecting Sea-Level Rise," *Science* 315 (2007): 368–70.

3. G. Helmar Gudmunsson, "Fortnightly Variations in the Flow Velocity of Rutford Ice Stream, West Antarctica," *Nature* 444 (2006): 1063–64.

4. Didier Swingedouw et al., "Effects of Land-Ice Melting and Associated Changes in the AMOC Result in Little Overall Impact on CO_2 Uptake," *Geophysical Research Letters* 34 (2007): L23706.

5. Peter Schwartz and Doug Randall, *An Abrupt Climate Change Scenario and Its Implications for United States National Security*, 2003, http://www.gbn.com /consulting/article_details.php?id=53.

Chapter 14

1. M. L. Parry et al., eds., *Contribution of Working Group II to the Fourth Assessment Report of the Intergovernmental Panel on Climate Change* (Cambridge: Cambridge University Press, 2007).

2. IPCC, Third Assessment Report, synthesis report, 2001.

3. Collectif Argos, *Les réfugiés climatiques*, preface by Hubert Reeves and Jean Jouzel (Gollion: Infolio, 2007).

4. IMPACTS, *Changements climatiques: Quels impacts pour la France*, Greenpeace-Climpact, preface by Jean Jouzel and Hervé Le Treut, 2005, http://www .impactsclimatiquesenfrance.fr.

5. M. L. Parry et al., eds., *Contribution of Working Group II to the Fourth Assessment Report of the Intergovernmental Panel on Climate Change* (Cambridge: Cambridge University Press, 2007).

6. Céline Le Bohec et al., "King Penguin Population Threatened by Southern Ocean Warming," *Proceedings of the National Academy of Sciences* 105 (2008): 2493–97.

Chapter 15

1. United Nations Framework Convention on Climate Change, FCCC/INFORMAL/84 GE.05-62220 (E) 200705, 1992, p. 5, http://unfccc.int/resource /docs/convkp/conveng.pdf.

2. http://www.globalcarbonproject.org/.

3. Bjorn Lomborg, *L'Écologiste sceptique: Le véritable état de la planète*, preface by Claude Allègre (Paris: Cherche-Midi, 2004).

4. Nicholas Stern, *The Stern Review Report: The Economics of Climate Change* (London: HM Treasure, 2006).

5. *Stratégie nationale d'adaptation au changement climatique* (Paris: Observatoire national sur les effets du réchauffement climatique, La Documentation française, 2007).

6. Pierre Radanne, *La division par quatre des émissions de dioxyde de carbone in France d'ici 2050* (report made to the MIES in March 2004); report writ-

ten under the presidency of Christian de Boissieu, *Division par quatre des emissions de gaz à effet de serre de la France à horizon 2050* (Paris: La Documentation française, 2006).

7. www.legrenelleenvironnement.fr

8. The detail of these measures as well as all the information pertaining to the Grenelle de l'environnement are found at http://www.legrenelle-environnement.fr/.

Chapter 16

1. Igor Zotikov, *The Antarctic Subglacial Lake Vostok* (Springer-Praxis, 2006).

2. G. de Q. Robin, D. J. Drewry, and D. T. Meldrum, "International Studies of Ice Sheet and Bedrock," *Philosophical Transactions of the Royal Society of London* 279 (1977): 185–96; A. Kapitsa, J. K. Ridley, G. de Q. Robin, M. J. Siegert, and I. Zotikov, "Large Deep Freshwater Lake beneath the Ice of Central East Antarctica," *Nature* 381 (1996): 684–86.

3. John Priscu et al., "Cosmicrobiology of Subglacial Ice above Lake Vostok, Antarctica," *Science* 286 (1999): 2141–44.

4. David Karl et al., "Microorganisms in the Accreted Ice of Lake Vostok, Antarctica," *Science* 286 (1999): 2144–47.

5. Sergy Bulat et al., "DNA Signature of Thermophilic Bacteria from the Aged Accretion Ice of Lake Vostok, Antarctica: Implications for Searching for Life in Extreme Icy Environments," *International Journal of Astrobiology* 3 (2004): 1–12.

Chapter 17

1. M. Murozumi et al., "Chemical Concentration of Pollutant Lead Aerosols, Terrestrial Dusts and Seasalts in Greenland and Antarctic Snow Strata," *Geochemica et cosmochemica acta* 33 (1969): 1247–48.

2. Claude Boutron et al., "L'archivage des activités humaines par les neiges et glaces polaires: Le cas du plomb," *Comptes Rendus Geoscience* 336 (2004): 847–67.

3. Jean-Pierre Candelone et al., "Post-industrial Revolution Changes in Large Scale Atmospheric Pollution of the Northern-Hemisphere by Heavy Metals as Documented in Central Greenland Snow and Ice," *Journal of Geophysical Research* 100 (1995): 16605–16.

4. Sungming Hong et al., "History of Ancient Copper Smelting Pollution during Roman and Medieval Times Recorded in Greenland Ice," *Science* 272 (1996): 246–49.

5. Hubertus Fischer et al., "Sulphate and Nitrate Firn Concentrations on the Greenland Ice Sheet 2, Temporal Anthropogenic Deposition Changes," *Journal of Geophysical Research* 103 (1998): 21935–42.

6. Michel Pourchet et al., "Some Meteorological Applications of Radioactive Fallout Measurements in Antarctic Snows," *Journal of Geophysical Research* 88 (1983): 6013–20.

7. Jean Jouzel et al., "A Continuous Record of Artificial Tritium Fallout at the South Pole (1954–1978)," *Earth and Planetary Science Letters* 45 (1979): 188–200.

8. Joseph Farman et al., "Large Losses of Total Ozone in Antarctica Reveal Seasonal CLO_x/NO_x Interaction," *Nature* 315 (1985): 207–10.

9. Paul Crutzen, "Geology of Mankind," *Nature* 415 (2002): 23.

Conclusion

1. Paul Crutzen, "Geology of Mankind," *Nature* 415 (2002): 23.

2. Private communication to the author (CL) at the French Academy of Sciences.

3. UN, *Global Environment Outlook: Environment for Development (GEO-4)* (2007).

SELECTED BIBLIOGRAPHY

Alley, Richard. *The Two-Mile Time Machine: Ice Cores, Abrupt Climate Change, and Our Future* (Princeton: Princeton University Press, 2002).

Collectif Argos. *Les réfugiés climatiques*. Gollion: Infolio, 2007.

Bard, Édouard, ed. *L'Homme face au climat*. Paris: Odile Jacob, 2006.

Berger, André. *Le climat de la Terre: Un passé pour quel avenir?* Brussels: De Boeck University, 1992.

Dansgaard, Willi. *Frozen Annals: Greenland Ice Sheet Research*. Odder, Denmark: Narayana Press, 2005.

Fellous, Jean-Louis, and Catherine Gautier, eds. *Comprendre le changement climatique*. Paris: Odile Jacob, 2007.

Imbert, Bertrand, and Claude Lorius. *Le Grand Défi des poles*. Paris: Gallimard, 2006.

IPCC. *Climate Change 2007: Fourth Assessment Report*. Cambridge: Cambridge University Press, 2007. Complete IPCC reports available in English only (Groups I, II, and III and synthesis report) on the IPCC site: http://www.ipcc.ch.

Joussaume, Sylvie. *Climat d'hier à demain*. Paris: CNRS Editions/CEA, 2000.

Le Treut, Hervé, and Jean-Marc Jancovici. *L'Effet de serre: Allons-nous changer le climat?* Paris: Flammarion, "Champs," 2004.

Lorius, Claude. *Glaces de l'Antarctique: Une mémoire, des passions*. Paris: Odile Jacob, 1991.

Metz, B., O. R. Davidson, P. R. Bosch, R. Dave, and L. A. Meyer, eds. *Contribution of Working Group III to the Fourth Assessment Report of the Intergovernmental Panel on Climate Change*. Cambridge: Cambridge University Press, 2007.

Pachauri, R. K., and A. Reisinger, eds. *Contribution of Working Groups I, II and III to the Fourth Assessment Report of the Intergovernmental Panel on Climate Change*. Geneva: IPCC, 2007.

Parry, M. L. Parry, O. F. Canziani, J. P. Palutikof, P. J. van der Linden, and C. E. Hanson, eds. *Contribution of Working Group II to the Fourth Assessment Re-*

port of the Intergovernmental Panel on Climate Change. Cambridge: Cambridge University Press, 2007.

Schneider, Stephen. "Science as a Contact Sport: Inside the Battle to Save the Earth's Climate," *National Geographic Society* (2009).

Solomon, S., D. Qin, M. Manning, Z. Chen, M. Marquis, K. B. Averyt, M. Tignor, and H. L. Miller, eds. *Contribution of Working Group I to the Fourth Assessment Report of the Intergovernmental Panel on Climate Change.* Cambridge: Cambridge University Press, 2007.

Weart, Spencer. *The Discovery of Global Warming.* http://www.aip.org/history.

Internet Sites

Institut Pierre-Simon-Laplace—http://www.ipsl.jussieu.fr

Intergovernmental Panel on Climate Change (IPCC)—http://www.ipcc.ch

Jean-Marc Jancovici—http://www.manicore.com

Laboratoire de Glaciology et Géophysique de l'Environnement (LGGE)—http://www-lgge.obs.ujf-grenoble.fr

Laboratoire des Sciences du Climat et de l'Environnement (LSCE)—http://www.lsce.ipsl.fr

Météo France—http://www.meteo.fr

Mission Interministérielle sur l'Effet de Serre (MIES)—http://www.effet-de-serre.gouv.fr

Organisation Météorologique Mondiale—http://www.wmo.c

INDEX

ablation, 20, 21–22, 29–30, 35, 69, 206
Adélie Land, 31, 34, 86, 98, 99, 250
Adhémar, Joseph Alphonse, 39, 49
aerosols, 72, 73, 115, 139, 170, 180–81, 183, 204, 262, 271; content of atmospheric aerosols, 116; effects of on clouds, 181; origins of, 181; radiative forcing of, 181; sulfur aerosols, 151
Africa, 44, 147, 149, 190, 222
Agassiz, Louis, 19, 37, 38
air bubbles, in the ice of Antarctica, 73–77, 103–4, 134, 143, 271; extraction of the bubbles from ice, 89–90; oxygen 18 found in, 119; relationship of nitrogen to oxygen in, 80, 126
air content, 75, 83, 116, 126, 191, 193. *See also* air bubbles, in the ice of Antarctica
Alaska, 9, 45
albedo, 7–8, 14, 118, 161, 176, 181, 196, 206–7
Aletsch glacier, 6
Alexander I, 31
alkenones, 62
Allègre, Claude, 189, 191
Allerod period, 130
Alps, the, 20, 49, 106, 187; effects of global warming on, 222–23; French Alps, 21, 207–8; Swiss Alps, 22–23
American Academy of Sciences, 191
American Geophysical Union, 153

Amery Ice Sheet, 13, 35
ammonium, 135
Amsterdam Island, 165
Amundsen, Roald, 24
Amundsen Sea, 26
Andes, 6, 23, 78, 223; glaciers of, 105–6
Angelis, Martine de, 90
Antarctic, 32
Antarctic Peninsula, 12, 13, 34–35, 55, 109, 199
Antarctic Treaty (1959), 85, 249
Antarctica, 7, 8, 10, 18, 38, 44, 68, 71, 72, 73, 85, 107, 114, 122, 141, 147, 212, 252, 253, 273; Australian research in, 104; chronology of ice in, 79; climate of, 123–24, 128–29; contribution of to the rise in sea levels, 212, 213; cooling of, 45; deep core drilling in central Antarctica, 105; drilling in by the Japanese, 103; exploration of by the French, 86–88; glaciation of, 45; ice sheets of, 5, 12–13, 247, 261; initial exploration of, 31–32, 34; movement of ice in, 69; research conducted by Europeans in, 100; separation of from Australia, 45; sulfates in, 266–67; uncertain mass balance of, 34–36; unexplored areas of, 18; warming of, 198–200. *See also* Dôme C; East Antarctica; West Antarctica
Anthropocene period, 247, 272–75

Arctic, the, 18; acceleration of global
     warming in, 197–98, 222; atmospheric
     pollution in, 262, 266–67; and the
     ozone hole, 271; and the problem of
     "Arctic fog," 266
Arctic Ocean, 8, 13, 225; exploration of,
     23–25; ice cover of, 196; projected
     warming of, 209–10; vulnerable ice of,
     25–27, 197–98
Argentina, 109
argon (Ar), 64, 65, 76, 143, 162, 175
Argos Collective, 225
Arolloa glacier, 208
Arrhenius, Svante, 39, 74–76, 89, 163,
     164, 174
ash layers, 78
Asia, 216, 222. See also Southeast Asia
astronomical forcing, 115
Atlantic Ocean, 148, 149, 215, 225. See
     also North Atlantic
atmospheric circulation, 137, 140
atomic mass, 54
Augustin, Laurent, 108
Australia, 44, 45, 122, 220, 231, 232, 235;
     increase in CO_2 emissions in, 233
Austria, 19, 22, 177

Baglety Ice Field (Alaska), 6
Bali Conference, 234–36, 238
Baltic Sea, 147
Bangladesh, 220
Bard, Édouard, 137
Barnola, Jean-Marc, 79
Belgium, 97, 100
Bellingshausen, Faddey, 31
Bellingshausen Sea, 26, 35
Bender, Michael, 80, 119
benthics, 61
Bentley Subglacial Trench, 12–13
Berger, André, 49, 65
Bering Strait, 24, 25
Bermuda, 60

beryllium 10 (^{10}Be), 73, 78–79, 128, 152
biodiversity, 177, 218, 227, 238, 241, 243,
     248, 251, 273
biomass, combustion of, 171
bison, 8–9
Bolivia, 23, 223
Bond, Gary, 143–44
Borchgrevink, Carstens, 32
Borloo, Jean-Louis, 243
Bossons Glacier, 19
Boutron, Claude, 263
Brazil, 147, 220; increase in CO_2
     emissions in, 233
British Antarctic Survey, 104
Broecker, Wally, 96, 131, 133, 134, 139,
     143
bromide (Br), 71, 170
Bronze Age, 19
Bruckner, Edouard, 39, 49
Brunhes, Bernhard, 126
Brunhes-Matuyama magnetic field, 65,
     126, 128
Bulat, Sergey, 257
Byrd, Richard, 32, 34
Byrd Station, 84–85, 89; ice drilling at,
     92, 147

Caillon, Nicolas, 191
calcium (Ca), 61, 62, 135
calcium carbonate ($CaCO_3$): formation of,
     60, 146; and isotopic fractionation, 61
"calibrating" curve, 64
California, 151
Callendar, Guy, 164, 173
Camp Century, 83–84, 87, 88, 89, 91, 107,
     109, 132, 173; ice drilling at, 92, 109,
     130, 131
Canada, 9, 149, 225, 232, 235, 249;
     northern Canada, 25
Canadian Arctic, 6, 37
Canon of Insolation and the Ice Age
     Problem (Milankovitch), 49

Capron, Emilie, 119
carbon (C), 54, 71, 204, 228; the carbon cycle, 204. *See also* carbon, isotopes of
carbon, isotopes of: carbon 12, 54; carbon 13, 54, 56, 59, 76, 153, 165; carbon 14, 54, 77–78, 79, 85, 128, 152; carbon 14 dating, 131; transformation of carbon 14 into nitrogen, 64
Carbon Budget 2009, 168
carbon dioxide (CO_2), 43, 122, 129, 133, 134, 153, 162, 175, 203, 204, 211, 262; absorption of by oceans and vegetation, 164–65, 168, 221; atmospheric CO_2, 75, 173–74, 175–76; CO_2 content of ice, 87; emissions of due to fossil fuels, 171–72; "heavy" CO_2, 56; seasonal variations in, 164–65; variations in the concentration of, 114–15, 125, 152–54, 164–66, 168, 191, 193, 204–5, 254; and volcanic activity, 44. *See also* global warming
carbon monoxide (CO), 170, 269
carbonates, 134
Caribbean, the, 216
Casey Base, 104
cellulose, 59
Chacaltaya Glacier, 23, 223
Champollion, Jean-François, 54
Chappellaz, Jérôme, 112, 147
Charcot, Jean-Baptiste, 32, 34, 86
Charcot base, 255
China, 23, 58, 63, 151, 216, 220; CO_2 emissions in, 171; increase in CO_2 emissions in, 233
chlorofluorocarbons (CFCs), 162, 170, 270, 271
chloride (Cl^-), 71, 170; chloride 36, 128
Chomette, Guy-Pierre, 225
chronology, 59, 110, 112, 126, 135, 137, 151, 154; of ice core samples, 78–79; of volcanic eruptions, 78

clathrates, 210; decomposition of, 44
Clean Development Mechanism (CDM), 232
climate, future of, 201; climate models concerning, 202–3; climatic projections derived from models, 205–6; the true threat of climate upheaval, 202–6. *See also* glaciers, future of
climate change: adaption to, 241–42; during the last millennium, 183–85; ice as an indicator of, 14, 16; importance of research concerning, 247–49. *See also* aerosols; climate/temperature oscillations; Earth, history of temperature and climate change on; greenhouse effect; greenhouse gases; human activity, awareness of the impact of human activity on the climate; human activity, and the composition of the atmosphere
Climate Conference. *See* United Nations Framework Convention on Climate Change (UNFCCC)
"Climategate," 189–90, 194
Climate/Long Range Investigation Mappings and Predictions Project (CLIMAP), 110
climate/temperature oscillations, 130, 149, 191, 193; and catastrophic events during warm periods, 139–42; clear indications of, 132–33; confirmation of, 134–37; connection of to the ocean, 143–44, 146; consequences of on a planetary scale, 147–48; initial evidence of, 91–92, 130–32; initial underestimating of temperature changes, 142–43
climatologists, 174–75
clouds, 87, 115, 116, 161, 175; cumulonimbus clouds, 4; effect of aerosols on, 181; and retroactions, 203–4
CNRS, 79

coccolites, 61
Cold Regions Research and Engineering Laboratory (CRREL), 83, 84, 89, 91
Commissariat à l'Energie Atomique (CEA), 83, 87
Concordia Station, 252, 258–60
Conference of Parties (COP), 230–31, 234–35
Congo, 23
continental archives, 62–64; dating of, 64–66
continental biosphere, 77, 115
Cook, James, 18, 31
Copenhagen, 82, 84, 136
Copenhagen conference, 236; failure of, 189, 238–39
copper (Cu), 264–66
coral/coral reefs, 60; of Barbados, 137; effect of global warming on, 221; formation of "terraces" on, 62; measurement of carbon 14 in, 64–65
cosmic rays, 152
Courtillot, Vincent, 189
Craig, Harmon, 96
Creseveur, Michel, 94
Croll, James, 39, 48, 76
Crutzen, Paul, 271, 274
cryosphere, 5, 14, 19, 199, 212, 248

Danish Istuk drill, 100
Dansgaard, Willi, 82, 83, 94, 96, 97, 98, 131, 132, 173
Dansgaard-Oeschger events, 92, 139, 142, 144, 148; and the analysis of stalagmites, 146, 147
Darwin glacier (Chile), 6
David, Hélène, 225
deforestation, 153, 168, 171, 238, 269
deglaciation, 116, 125, 136, 148, 149
Delmas, Robert, 89, 90, 111
dendroclimatology, 59, 60
Denmark, 28, 97, 100, 225, 249

deuterium (D), 54, 55, 70, 76, 85, 120, 128; analysis of on the ice of Antarctica and Greenland, 90, 146
diatoms, 61
Dôme A, 260
Dôme C, 97, 100, 103, 111, 119, 122–23, 148, 155, 165, 253, 254, 259; dating of ice cores in, 78, 126; drilling to reach the bedrock of, 98–99; ice core drilling in, 88–90, 92, 108, 124–26; interest of astronomers and astrophysicists in, 260
Dôme Fuji, 103, 108, 109, 123, 126
Donnadieu, Yannick, 43
Donnou, Daniel, 89
Drake Passage, 45
Drewry, David, 100
Dreyfus, Gabrielle, 119, 126
Drobriansky, Paula, 235
Dronning Maud Land (DML), 100, 147–48
Dubois, Jacques, 86
Dumont d'Urville, Jules, 7, 31, 86
Dumont d'Urville Base, 86, 88, 99, 100, 258; average temperature recorded at, 7
dust, 71, 72, 78, 90, 111, 115, 135, 137, 144, 254, 263; fallout of desert dust, 119; magnetic dust, 42; radioactive dust, 268
Duval, Paul, 90
Dye3, 92, 110, 111, 132, 134, 135, 137; changes recorded at, 143

Earth, 40, 75, 116, 203, 248; absorption of the Sun's energy by, 159–60; axis of rotation of, 7, 159; connection between Earth's position and its orbit around the Sun, 39; inversion of the Earth's magnetic field, 126, 128–29; magnetic field of, 261–62; orbit of and glacial periods, 46–47, 79, 173;

"Snowball Earth," 43; surface temperature of, 40, 161–62; water content of, 3. *See also* Earth, history of temperature and climate change on

Earth, history of temperature and climate change on: and the accumulation of data from oceanic and continental archives, 66–67; and the distant past, 60; limitations of the historical approach to climate measurement, 58–59; and the recent period, 57–60. *See also* continental archives; human activity, and the composition of the atmosphere; loess, climatic indications present in

Earth Summit, 227

East Africa, 23

East Antarctica, 12, 13, 34–35, 87, 100, 103, 123, 199, 255, 256

East Germany, 231

Eastern Europe, 231

Échets, 146

Eemian period, rapid/catastrophic events of, 139–42

Egypt, 58

Ekström, Göran, 198

El Niño, 105, 148, 150, 185

El Salvador, CO_2 emissions in, 171

Elephant Island, 27

Elkohlm, Nils, 75

emperor penguins, 250–51

Endurance, 26, 32

England. *See* United Kingdom

Ente per le Nuove Technologie l'Energia e l'Ambiante (ENEA), 99

Eocene/Oligocene transition, 44–45

EPICA (European Project for Ice Coring in Antarctica), 100, 108–9, 122, 126, 154, 259

Erebus, 24

Eric the Red, 28

Eurasia, 45

Europa, 257

Europe, 152, 220, 235; reduction of greenhouse gas emissions in, 233. *See also* Eastern Europe; Western Europe

European Community (EC), 231, 233

European "Great Challenges," 100

European Project for Ice Coring in Antarctica. *See* EPICA (European Project for Ice Coring in Antarctica)

European Remote Sensing (ERS 1) satellite, 30–31, 256

European Science Foundation (ESF), 97, 100

European Space Agency (ESA), 31

evaporation, 94, 174, 176, 202, 222; and "latent heat," 163; of ocean water, 16, 60, 62, 163, 174, 176

Expéditions Polaires Françaises (EPF), 86, 99

Filchner Shelf, 34

Finland, 249

fission, 65

Florida, 221

fluoride (F), 71

foraminifera, 61; concentration of chemical elements in, 62; concentration of oxygen 18 in, 61–62; marine foraminifera, 131

fossil fuels, 164, 165, 168, 178, 181, 210, 228, 233, 237, 264, 269, 272; CO_2 emissions from, 171–72, 204

Foucart, Stéphane, 191

Fourier, Joseph, 75, 76, 163

fractionation: of gaseous compounds, 143; isotopic fractionation, 55–56, 76; and precipitation, 61; water and the fractionation process, 55–57. *See also* calcium carbonate ($CaCO_3$), and isotopic fractionation

Fram, 24

Français, 32

France, 22, 63, 97, 100, 101, 121, 189, 231,
    255; effects of global warming in,
    222–23; generation of electricity in,
    172; reduction of greenhouse gases in,
    233 (*see also* Grenelle de
    l'environnement); temperature record
    of, 58
Franklin, John, 23–24
Franz Josef Land, 25
Franz-Josef glacier, 208
French Académie des Sciences, 195

Gauss, 32
Gébroulaz glacier, 208
Gendrin, Roger, 101
General Circulation Models, develop-
    ment of, 202
geodesic chaining, 29
geophysicists, and the "inverse" methods,
    80
geothermal flux, 30
Gerlache, Adrien de, 32
Germany, 60, 97, 99, 100; reduction of
    greenhouse gases in, 233
Gif-sur-Yvette, 111
Gillet, François, 89
glacial archives, 68; and air bubbles,
    74–77; multiple sources of impurities
    in, 71–73
glacial lakes, 102, 120–21, 149
glacial periods, 131–32
glacial-interglacial cycle, 112, 149, 154–55,
    253
glaciations, 123, 253, 254; alignment of
    glacial deposits, 42–43; in Antarctica,
    45; and atmospheric CO_2, 39; in
    Europe (the Wurm, Riss, Mindel, and
    Gunz glaciations), 39, 41; magnetic
    dust in glacial deposits, 42; Ordovi-
    cian glaciation, 44; past glaciations,
    41–46; Permian glaciation, 44; of the
    Quaternary and astronomic theory,

46–49, 163; theories of past glacia-
    tions, 38–39, 41
Glacier de Leschaux, 20
Glacier de Saint-Sorlin, 21
Glacier de Talèfre, 20
Glacier des Périades, 20
Glacier du Géant, 20
glaciers, 16, 18, 197–98, 200; Alpine
    glaciers, 6, 21, 23, 88, 180, 206–7; area
    covered by mountain glaciers, 6;
    dynamics of, 19–20; decrease in
    albedo of, 206–7; future of, 206–9;
    loss of through ablation, 206; mass
    balance and glacier health, 21–23;
    mountain glaciers, 5–7; number of
    mountain glaciers, 6; position of the
    terminal tongue (front of a glacier),
    20–21; variations in length of, 20–21.
    *See also specific individually listed
    glaciers*
glaciochemists, 72
glaciological models, 79–80
glaciologists, 68, 70
Global Carbon Project, 168, 233
Global Change program, 177
global warming, 116, 179–80, 273–74;
    acceleration of, 185–86, 197–98;
    certainty of, 185–87; differences
    between warming measured on Earth's
    surface and the atmosphere, 194;
    skepticism concerning, 189–91,
    193–95; and the white planet,
    195–200. *See also* global warming,
    multiple consequences of; global
    warming, solutions for
global warming, multiple consequences
    of, 218; for agriculture, 219; for
    animals, 223–24; for coastal areas,
    220; for coral, 221; economic and
    political consequences, 225–26; and
    general global upheaval, 218–22; for
    human health, 221–22; for human

populations in the Arctic, 224–25; for
mountain ranges, 222–23; and the
possibility of "climate refugees,"
220–21, 225; for polar ecosystems,
223–25; for tourism, 221, 222, 223,
248; water resource problems, 219–20
global warming, solutions for, 227;
meeting the challenges of slowing
global warming, 236–37, 248;
stabilizing the greenhouse effect,
228–30. *See also* Bali Conference;
Grenelle de l'environnement; Kyoto
Protocol
Gondwana, 44
Gorbachev, Mikhail, 250
Gore, Al, 209–10, 221, 234, 275
GRACE (Gravity Recovery and Climate
Experiment), 31, 36
grape harvests, and the recording of
summer temperatures, 58
gravity, 76
Gravity Recovery and Climate Experiment. *See* GRACE (Gravity Recovery
and Climate Experiment)
greenhouse effect, 74–75, 129, 176, 183,
203, 250; as a beneficial natural
phenomenon, 159–63; and climate,
113–16, 118; as a result of human
activity, 163–66, 168–72; stabilization
of, 228–30
greenhouse gases, 43–44, 116, 162, 203,
221, 250, 254, 269, 273; anthropogenic greenhouse gas concentrations,
185, 187, 271; "indirect" greenhouse
gases, 172; reduction of in Europe,
233; stabilization of, 227
Greenland, 7, 12, 18, 25, 35, 38, 49, 68, 71,
72, 73, 85, 125, 130, 132, 141, 180, 194,
221; accumulation and ablation
conditions of, 29–30; atmospheric
pollution of, 264–65; climate
relationships between Greenland and

the North Atlantic, 144–45; coastal
regions of, 197–98; contribution of to
the rise in sea levels, 212, 213, 214;
dating of ice cores in, 78; first
glaciological measurements of, 28–29;
highest region of the Greenland
Plateau, 97; ice sheets of, 5, 10–11, 30,
46, 82–83, 200, 247, 253, 260;
impurities in the ice of, 134; inhabita-
tion of, 28; isotopic records of, 111;
lead pollution in, 262–64; methane in
the ice of, 149–50; mini-glacial
earthquakes in, 198; movement of ice
in, 69; and negative mass balance,
28–31; "sawtooth" sequences/
structure of Greenland ice, 137, 144,
148; size of, 10; subglacial river
discovered under the ice of, 107;
sulfates in, 266–67; surface of covered
by ice, 11; temperature measurements,
76, 146, 147, 148, 149; temperature
variations in the ice of, 136–37, 149.
See also Greenland, ice drilling in by
researchers from Europe and the
United States
Greenland, ice drilling in by researchers
from Europe and the United States,
96–98, 106–7. *See also* GRIP
(Greenland Ice Core Project); GISP
(Greenland Ice Sheet Project
Greenland Ice Core Project (GRIP). *See*
GRIP (Greenland Ice Core Project)
Greenland Ice Sheet Project. *See* (GISP)
Greenland Ice Sheet Project
Greenland Sea, 25
Grenelle de l'environnement, 242–44,
274
Grenoble, 79, 88, 89, 90, 94, 98, 104, 108,
119, 128
GRIP (Greenland Ice Core Project), 97,
100, 103, 106, 125, 134, 147; dating of
the GRIP ice cores, 135–37;

GRIP cont.
  instabilities in the GRIP records,
    140–41; North GRIP, 107 109, 146,
    253; origins of, 97–98; success of, 99;
    work of in the "scientific trench,"
    135–36
GISP (Greenland Ice Sheet Project), 91;
    GISP2, 97, 98, 106, 125, 134, 137, 141
growth rings, of trees. *See*
    dendroclimatology
Gulf Stream, 12, 134, 144; possible halting
    of, 214–17
Gunderstrup, Niels, 91

Haiti, CO_2 emissions in, 171
Halligen Islands, 220
halogen, 71
halogen compounds, 162, 168–69,
    170–71, 172
Hansen, Jim, 176, 185–86, 193–94, 203
Hayes, Jim, 49, 65, 90, 110
Heinrich layers, 143, 144
heliothermometer, 163
Highjump operation, 32–33
Himalayas, 6, 105
Högbom, Arvid, 75
Hoggar, 44
Holocene period, 66–67, 104, 132, 149,
    150, 166, 272; monsoons during, 150;
    stability of the climate during,
    123–24; volcanic eruptions during, 78.
    See also human activity, and the
    composition of the atmosphere
Homo erectus, 272
Homo habilis, 272
Howard, John, 232
Huet, Sylvestre, 191
human activity, 272–73; awareness of the
    impact of human activity on the
    climate, 174–76, 201; and the
    composition of the atmosphere,
    152–55, 168–72, 173–74; and the rise

of pollution, 261–62; and temperature
    changes in the Northern Hemisphere,
    183–85
Humboldt Glacier, 11–12
hydrocarbons, 172
hydrochlorofluorocarbons (HCFCs),
    170–71
hydrofluorocarbons (HFCs), 171
hydrosphere, 3, 3–5, 8, 257; as an agent
    and indicator of climate change, 14,
    16; air bubbles in, 74–77; blending of
    ice layers and the dating of ice cores,
    141–42; chemical composition of
    modified by volcanic activity, 151;
    CO_2 content of, 87; decrease in sea
    ice, 176; electrical conductivity of,
    136; ice older than that at Vostok,
    123–26; isotopes of, 70–71; issues
    concerning ice dating, 77–81;
    microbiology of, 255–58; plasticity of,
    19, 69; "sawtooth" sequences/structure
    of Greenland ice, 137, 144, 148; sea ice,
    5, 12, 16, 24–25, 31, 32, 62, 78, 83, 99,
    115, 116, 150, 176, 196–97, 198–99,
    201, 223–24, 248 (*see also* Arctic
    Ocean); transition between water
    vapor and ice, 16

ice caps, 5–7, 11, 14, 16, 18, 37, 38, 70, 88,
    158, 200, 201, 219, 212, 248, 254;
    altitude of glacial ice caps, 77; glacial
    ice caps, 6–7, 10; in Iceland, 6; total
    area of, 6–7
ice core drilling, 110–11; and the dating of
    ice cores, 78–81; deep core drilling in
    central Antarctica, 103–5; deviations
    in drilling of ice cores, 92; drilling by
    the Japanese, 103; drilling in Green-
    land, 106–7; initial deep ice core
    drillings, 82–85; objectives of glacial
    ice coring, 253–54; problems with
    drilling for ice cores, 141. *See also*

Dôme C; European Project for Ice
Coring in Antarctica (EPICA);
Greenland, ice drilling in by researchers from Europe and the United
States; Vostok/Vostok Station
ice platforms, 10–14
ice sheets, 14, 16, 18, 30, 38, 40, 42, 44,
62, 70, 77, 86, 112, 118, 151, 187, 191,
272, 274; of Antarctica, 5, 12–13, 247,
260; density of, 69; depletion of
oxygen 18 in, 61; of Greenland, 5,
10–11, 30, 46, 82–83, 200, 247, 253,
260; loss of, 200; of North America,
144, 149; of the Northern Hemisphere, 46, 122–23
ice shelves, 11, 199–200, 252; Larsen B Ice
Shelf, 199; of West Antarctica, 13–14;
Wilkins Ice Shelf, 199
icebergs, 12, 13–14, 144; dating of
Greenland icebergs, 83; influence of
on deep water circulation, 66
Iceland, 6, 97, 250
Imbrie, John, 37–39, 49, 90, 110
Imbrie, Katherine, 38–39
India, 150, 220; increase in CO_2 emissions
in, 233
Indian Ocean, 147, 165; dating of cores
from, 65
infrared radiation, absorption of by the
atmosphere, 162
inlandis ("ice in the middle of the land"),
10
insolation, 77, 118, 150, 159–60; measure
of by astronomical parameters
(eccentricity, obliquity/tilt, precession
of the equinoxes), 46–48, 160
Institut de Recherche et Développement
(IRD), 106
Institut Français de Recherche et de
Technologies Polaires (IFRTP), 99,
100, 101, 104, 252
Institut Pierre-Simon-Laplace, 202

Institut Polaire Français Paul Emile
(IPEV), 104, 252, 254, 259
Institute of Microbiology of the Academy
of Sciences (Russia), 256
InterAcademy Council, 190
Intergovernmental Panel on Climate
Change. *See* IPCC (Intergovernmen-
tal Panel on Climate Change)
International Council of Scientific
Unions (ICSU), 177, 249, 251
International Energy Agency (IEA), 230
International Geophysical Year (IGY
[1957–1958]), 29, 34, 85, 86, 249, 255,
261; launching of satellites during, 250
International Geosphere-Biosphere
Program (IGBP), 177
International Glaciological Expedition,
86
International Partnership in Ice Core
Sciences (IPICS), 253
International Polar Year (IPY), 226, 247,
251–53; contribution of the French to,
252–53
International Union for the Conservation
of Nature (IUCN), 224
inverse method, 80
inversion, of the Earth's magnetic field,
126, 128–29
Inybtehek glacier (Tian Chan), 6
iodine (I), 71
IPCC (Intergovernmental Panel on
Climate Change), 168, 173, 185, 187,
193, 201, 209, 220, 221, 228, 229, 231,
235, 247, 275; "errors" of, 190–91;
establishment of, 177–80; on the
possibility of reducing greenhouse
gases, 236–37; report of on aerosols,
180–81; report of on human activity
and climate change, 180; report of on
the rise of sea levels, 212–13;
skepticism concerning data used by,
189–91

"iris" effect, 203–4

isotopes, 55, 125; chemical properties of, 55–56; cosmogenic isotopes, 73, 78, 152; isotopic analysis, 63; isotopic geochemistry, 55; radioactive isotopes, 55. *See also* fractionation, isotopic; isotopic content; isotopic thermometer

isotopic content: of ice, 140; of snow, 70, 85, 135; of water, 56, 57, 60, 63

isotopic thermometer, 55; skewing of, 70–71

Italy, 97, 100; reduction of greenhouse gases in, 233

Jakobshavn glacier, 12, 30

James Ross Island, 109

Japan, 235

Jeannette, 24

Jensen, Dorthe Dahl, 253

Johansen, Hjalmar, 24–25

Johnson, Sigfus, 91, 132

Jones, Phil, 189, 193–94

Jouzel, Jean, 87–88, 94, 96, 100, 189, 191, 234, 243, 252

Jugie, Gérard, 104

Juppé, Alain, 242–43

Karl, David, 257

Keeling, Charles, 74, 164–65; and the "Keeling curve," 164

Kennett, James, 45

Kepler, Johann, 46

Kilimanjaro glacier, 23

Kirschvink, Joseph, 43

Kohnen Station, 100

Köppen, Wladimir, 48–49, 174

Kotlyakov, Volodya, 94

Kyoto Protocol, 230–34, 239

La Niña, 185, 195

Labeyrie, Jacques, 88

Labeyrie, Laurent, 119

Laboratoire de Géochimie Isotopique (LGI), 87

Laboratoire de Glaciologie et de Géophysique de l'Environnement (LGGE), 88, 111, 112; LGGE Grenoble, 106

Laboratoire des Sciences du Climat et de l'Environnement (LSCE) Saclay, 106, 119, 165

Ladurie, Emmanuel Le Roy, 58

Lake Agassiz, 149

Lake Ojibway, 149

lake sediments, 53, 60, 63; and the measurement of climate change, 61

Lake Vostok, 102, 120, 252, 255–56, 257, 258

Lambert, Gérard, 165, 210

Lambert glacier, 13; surface area of, 35

Landais, Amaëlle, 119, 143

Langway, Chet, 84, 96, 131

Larsen Ice Shelf, 35; Larsen B Ice Shelf, 199

Laskar, Jacques, 49

Last Glacial Maximum, 76, 87, 104, 110, 111, 116, 131, 142, 203, 238

Laurentide ice sheet, 71

lava, analysis of, 126

Law Dome, 104, 109, 166

Le methane et le destin de la Terre. Les hydrates de methane: Rêve ou cauchemar? (Lambert), 210

Le Monde, 191

lead (Pb), 262–64

Legrand, Michel, 90

Leningrad Arctic and Antarctic Research Institute, 93–94

Leningrad Mining Institute, 92; interest of in drilling techniques, 93

Les glaces de l' Antarctique (Lorius), 86, 92

Libération, 191

L'imposteur c'est lui (Huet), 191

Lindzen, Richard, 203
Lipenkov, Volodya, 101, 102, 103
Little Ice Age, 58, 152, 180, 183, 194
Lliboutry, Louis, 88
loess, 60; climatic indications present in, 63–64
Loewe, Fritz, 28
Lorius, Claude, 74, 86, 87, 88, 94, 96, 100, 252
Lyell, Charles, 40

Ma vérité sur la planète (Allègre), 189
Madrid Protocol, 226
magnesium (Mg), 60, 62
Malaizé, Bruno, 119
Malaspina glacier (Alaska), 6
Maldive Islands, 220
mammoths, 8–9, 210
Marie Byrd Land, 12
marine biosphere, 221
Marine Isotopic Stage (MIS), 153
marine salt, 72
marine sediments, 49, 63, 64, 66, 90, 96, 110, 111, 128, 131, 143, 144, 147, 153, 155, 253; composition of, 61
Mars, 203, 257
Massif Central caves, 146, 187
Masson-Delmotte, Valérie, 146, 253
Mauna Loa, 164
Maunder Minimum, 152, 160
Mayewski, Paul, 96
McMurdo Base, 84, 85, 89, 104
Mediterranean area, 218, 220
Mediterranean Sea, 147
Mer de Glace glacier, 6, 20, 37–38, 209
Mercer, John, 14
Météo France, 202
methane (CH4), 43, 76, 114, 162, 168–69, 172, 175, 269, 271; formation of, 210–11; hydrates of, 210; increase of during the Paleocene-Eocene Thermal Maximum, 44; oxidation of, 170; production of, 169–70; production of and the extent of flooded zones, 147; thermal methane, 211; variations in methane concentration, 115, 125, 152, 153–54
methanogenic bacteria, 211
microbiologists, 121
Milankovitch, Milutin, 39, 48–49, 90, 112
Miller, Heinz, 100
Mirney Station, 92
Moisselin, Jean-Marc, 58
Monaco glacier (Spitzberg), 6
Monod, Théodore, 275
monsoons, 66, 105, 199, 126, 147, 149, 150, 203
Mont Blanc, 19, 21
Mont Blanc du Tacul, 20
Mont-Blanc Massif, 222
Montreal Protocol, 271
Moon, the, 257
Morel, Pierre, 177
Moscow Institute of Geography, 94
Mount Kirkpatrick, 13
mountain massifs, 38
Mullen, George, 40
Muller, Richard, 193
Muztagh Ata, ice core drilling by the Chinese at, 106

Nansen, Fridtjof, 24–25, 250
National Aeronautic and Space Administration (NASA), 30, 176, 202
National Center for Atmospheric Research (NCAR), 202, 209
National Science Foundation (NSF), 89, 96, 101
Nature, 94, 110, 114, 119, 140
neodymium (Nd), 72
Neo-Proterozoic period, 42
Netherlands, the, 100, 190, 216; freezing of canals in, 58

New Guinea, 23

New York Herald, 24

New York Times, 94

New Zealand, 208

Ngojumba glacier (Nepal), 6

Nigardsreen glacier, 208

nitrate, 135

nitrogen (N), 54, 64, 71, 76, 77, 162, 165, 175

nitrogen isotopes, 88

nitrogen oxides (NO_x), 172, 269

nitrogen protoxide (N_2O), 76, 172

nitrous oxide (N_2O), 162, 168, 171, 172

Nordenskjold, Nils, 24

North Africa, 150

North America, 23, 45, 223; droughts in, 150

North Atlantic, 12, 25, 132, 133, 134, 144, 148, 216

North Pole, 7, 8, 18, 37, 225–26; international observatories of, 249–50

North Sea, 220

Northern Hemisphere, 5, 9, 42, 48, 112, 115, 146, 147, 149, 150, 154, 165, 198, 215, 261, 270; amount of ice in, 38; history of pollution in (1773–1992), 264–66; ice sheets of, 46, 122–23; monsoons of, 119; nuclear testing in, 73; radioactivity in, 268; snow-covered surfaces of, 196; theories for the presence of ice in, 38–39, 41; warming of, 116, 118, 183–84

Northwest Passage, 24

Norway, 28, 100, 208, 225, 250

Norwegian Sea, 27, 133

nuclear energy, 84, 237

nuclear technology, 99

nuclear weapons, 216; testing of, 73, 78, 85, 268

nunataks ("land in the middle of glaciers"), 12

Observatoire National sur les Effets du Réchauffement Climatique (ONERC), 241

oceanic archives, dating of, 64–66

oceans, 56; and the absorption of CO_2, 164–65, 168, 221; CO_2 content of the oceans, 133; evaporation of ocean water, 16, 60, 62, 163, 174, 176; frozen oceans, 43; ocean circulation, 133–34; salinity of, 62; thermal expansion of, 187, 194

Oerlemans, Hans, 21

Oeschger, Hans, 74, 85, 89, 96, 97, 132, 133, 143

"On the Influence of Carbonic Acid in the Air upon the Temperature of the Ground" (Arrhenius), 74–75

orbital forcing, 114

Ordovician period, 44

oxidation, 71

oxygen (O), isotopes of: analysis of oxygen 18 on the ice of Antarctica and Greenland, 90; concentration of oxygen 18 in surface ocean waters, 62; oxygen 17, 54; oxygen 18, 54, 55, 56, 59, 61, 63, 70, 76, 80, 85, 120, 131, 136; oxygen 18 and the composition of precipitate carbonate, 60; oxygen 18 in air bubbles, 119; oxygen in seawater, 144

ozone (O_3), 162, 172, 221; ozone layer, 171, 178; stratospheric ozone, 178

ozone hole, 269–71

Pachauri, Rejendra, 234

Pacific Ocean, 133, 147, 225

paleoaltimeter, 77

Paleocene-Eocene Thermal Maximum, 44

paleoceanography, 61–62

paleoclimates, 49

paleoclimatologists, 45, 57, 120

paleoclimatology, 60; guiding principle of, 54–55

paleomagneticians, 42
palynology, 63
Parc Montsouris observatory, 270
Paris, 47–48
Parrenin, Frédéric, 80
Patagonia, 23, 119
Patterson, Clair, 262–63
Peltier, Dick, 65
Penck, Albrecht, 39, 49
Pensacola Mountains, 34
perfluoride compounds (PFCs), 171
permafrost, 5, 18, 83, 198, 210–11, 225, 248, 252; permafrost zones, 9–10
Permian period, 44
Petit, Jean-Robert, 90, 94, 101, 119
photosynthesis, 57
Pic du Midi observatory, 270
Pierrehumbert, Ray, 191
plankton, 61
polar amplification, 195–96
polar bears, decreasing habitat of, 224
pollen, 63, 130, 131, 146
pollution: from heavy metals including copper, 264–66; history of in the Northern Hemisphere (1773–1992), 264–66; lead pollution, 262–64; from radioactivity, 268–69; rise in due to human activity, 261–62; spread of in the Arctic and Antarctic, 262; from sulfates, 266–67. *See also* ozone hole
polynya (open water surrounded by sea ice), 26
Portugal, reduction of greenhouse gases in, 233
potassium 40 (K), 64
Pouiller, Claude, 163
Pourquoi Pas, 32
Precambrian period, 40
precipitation, 79, 174. *See also* precipitation, types of
precipitation, types of: graupels, 4, 5; hail, 4, 5; snow, 4–5

Princeton Geophysical Fluid Dynamics Laboratory (GFDL), 202
Priscu, John, 257
Putin, Vladimir, 232
Pyrenees, 222

Qin-Dahe, 185
Quaternary period, 40, 45–46, 63, 65, 66, 90, 110, 160, 250, 274; glaciations of, 46–49, 163
Queen Maud Land, 34
Quervain, Alfred de, 28

radioactivity, 268–69
radiolars, 61
Raisbeck, Grant, 79, 119, 128
Randall, Doug, 216
Raynaud, Dominique, 87, 89, 119, 189, 252
RealClimate, 191
"Remarques générals sur les températures de globe terrestre et des espaces planétaires" (Fourier), 75
Resolute Bay, 7
Revelle, Roger, 164
Rignot, Éric, 198
Robin, Gordon, 256
Rodinia, 43
Romania, reduction of greenhouse gases in, 233
Ronne-Filchner Ice Shelf, 13, 14
Ross, James, 31–32
Ross Ice Shelf, 13, 14, 34
Ross Sea, 26, 34
Roth, Étienne, 83
rubidium (Rb), 72
Rudd, Kevin, 232
Ruddiman, Bill, 153–55
Russia, 9, 101, 121, 225, 231, 232, 235, 250, 255; importance of snow to, 10–11; reduction of greenhouse gases in, 233
Rutford glacier, 213

Rwenzori Mountains, 23

Sagan, Carl, 40
Sahara Desert, 44
Saint Petersburg Mining Institute, 101
Saint-Sorlin glacier, 208
Santorini, volcanic eruption on, 151
Sarkozy, Nicolas, 242, 243
"sawtooth" sequences/structure, 137, 144
sea levels, 16–17; rapid rise of, 211–14
Scandinavia, 23, 85
Schlich, Roland, 86
Schmidt, Gavin, 191
Scholander, Per, 73
Schwartz, Peter, 216
Science, 110; letters and articles concerning scientific ethics in, 191
Scientific Committee for Antarctic Research (SCAR), 249
Scott, Robert, 32
Scott Polar Research Institute, 255
Severinghaus, Jeff, 142
Shackleton, Ernest, 26, 32
Shackleton, Nick, 45, 49, 65, 90, 110
Shackleton Mountains, 34
Shishmaref, 225
Siachen glacier (Karakorum), 6
Siberia, 9, 10, 23, 216
Siberian Arctic, 37
Sierra Nevada Mountains, 223
silica, 61
Siple Dome, 105
snow, 3–5, 55, 73, 268, 269; density of, 5–6; "dry" snow, 12; importance of to Russia, 10; isotopic content of, 70–71; life expectancy of snowflakes, 68–69; *névé* snow, 5–6, 72; permanent snow, 5; snow shapes in Siberia (*zastrougi* and *sastrugi*), 10
solar constant, 160
solar radiation, 46, 161, 175, 176, 181, 196
Solomon, Susan, 185, 221
South Africa, 220

South America, 44, 45, 119, 122, 223; climate change in, 150
South Georgia, 32
South Pole, 11, 31, 32, 37, 44, 104, 164, 255; isotopic content of snow at, 85
South Pole Station, 164
Southeast Asia, 150, 220
Southern Hemisphere, 5, 42, 147, 148, 165, 261
Southern Ocean, 8, 45, 118
Sowers, Todd, 119
Spain, reduction of greenhouse gases in, 233
Spitzberg, 25
stalagmites, 63; and Dansgaard-Oeschger events, 146, 147
Stauffer, Bernhard, 133
Steffensen, Jorgen Peder, 146
Stern, Nicholas, 243; and the Stern Report, 237
Stockholm Physics Society, 75
stratosphere, 162, 170, 171, 172, 262, 268, 270; cooling of, 194, 195, 271; thermal structure of, 269. *See also* ozone hole
strontium (Sr), 60
subduction, 62
subglacial floors, topography of, 13
subglacial lakes, 121, 149, 255; discovery of a subglacial lake below Vostok Station, 102, 120–21, 252, 255–56, 258; microbiology of, 255–58
subglacial ocean, 257
subglacial river, 107
Subsidiary Body for Scientific and Technological Advice (SBSTA), 231
Suess, Hans, 164
sulfates, 78, 266–67
sulfur (S), 71, 72, 151
Sullivan, Walter, 94
Summit Station, 97, 98, 264, 266–67
Sun, the, 7, 20, 38–39, 75, 159–60, 174; beams of, 7–8; increase in the luminosity of, 40; luminosity of, 43,

44; paradox of the pale Sun, 40; solar activity and volcanism, 150–52; surface temperature of, 161; variations in solar activity, 160–61. *See also* solar radiation
Svalbard ice cap (Spitzberg), 6
Sweden, 100, 250
Swedish Meteorology Office, 75
Swiss Clariden Glacier, 21
Switzerland, 22, 97, 100

Talos Dome, 104, 254
Tasman glacier (New Zealand), 6
Taylor Dome, 104
temperature, 3–4, 5, 6, 8, 14, 16, 55, 58–60, 62, 68, 75, 76, 90, 124, 125, 128, 144, 150, 155; average temperature of the continents, 174, 176, 177; of celestial bodies, 161; evolution of mean temperature of the Earth, 183, 185; and the formation of glacial lakes, 149; and the fusion point, 69; increase in ocean temperatures, 187; and the isotopic content of snow, 70, 135; measurement of average global temperature, 193; pre-industrial temperature, 230; temperature curve of the Earth's surface, 40, 44–45; temperature record in France, 58; temperature record in Greenland, 76, 146, 147, 148, 149; underestimated changes in, 142–43. *See also* climate/ temperature oscillations; Earth, surface temperature of; global warming; greenhouse effect, stabilization of; Sun, the, surface temperature of
Terra Nova Bay, 99, 100, 104, 258, 259
Territoires des Terres Australes et Antarctique Françaises (TAAF), 99
Terror, 24
thermal expansion, inertia of, 213
thermohaline circulation, 148, 149;

halting of, 215–16; inverse thermohaline circulation, 215
thermophilic bacteria, 257
Thompson, Lonnie, 105–6
Tiros satellite, 30
Toba, 78
Transantarctic Mountains, 12
trees, and the measurement of climate history, 59–60
Treut, Hervé Le, 203
tritium (^{3}H), 85, 268–69
troposphere, 162, 195, 221, 269
Tuvalu Islands, 220
Tyndall, John, 163
Tyrol, the, 19

Uganda, 23
ultraviolet radiation, 270
United Kingdom, 97, 99, 100; reduction of greenhouse gases in, 233
United Nations Environment Programme (UNEP), 178
United Nations Framework Convention on Climate Change (UNFCCC), 227, 230, 238, 239
United States, 101, 121, 147, 216, 220, 225, 231, 232, 249, 255; CO_2 emissions in, 171, 233; opposition of to the data of the Bali Conference, 235–36; use of unleaded gas in, 264
Uppsala (Patagonia), 6
uranium-thorium dating method, 65
urbanization, 58, 139
Urey, Harold, 61
U.S. Army Corps of Engineers, 83

Vallée Blanche, 20
Van Allen Belt, 250
Vasiliev, Nikolay, 102
Vatnajokull ice cap (Iceland), 6
Venus, 203
Victor, Paul-Émile, 28, 34
Victoria Land, 31, 86

Vietnam, CO_2 emissions in, 171
Villach Conference (1985), 177
Villars cave, 63, 146
Vimeux, Françoise, 119
Vincent, Christian, 20
volcanoes/volcanic activity: during the Holocene period, 78; effects of volcanic eruptions on the climate, 151–52; and increases in the greenhouse effect, 151; lava analysis, 126; volcanic eruptions, 71–72; volcanism and solar activity, 150–52
Vosges Mountains, 63; Grand Pile of, 146
Vostok/Vostok Station, 7, 108, 110, 112, 125, 126, 153, 154, 165, 176, 253, 254, 255; CO_2 concentrations at, 114–15; destruction of the generator at, 101; discovery of a subglacial lake below Vostok Station, 102, 120–21, 252, 255–56, 258; information gathered from ice cores of, 118–19; initial ice drilling at by French and Soviet teams, 92–94, 96, 101–3; mean annual temperature of, 112; variations in temperature at, 114, 115–16; warming of, 118. *See also* Lake Vostok

water (H_2O), 170, 210, 257; chemical composition of, 55; energy redistribution and the water cycle, 163; evaporation of, 16, 60, 163; and the fractionation process, 55–57; freezing point of seawater, 3; freshwater, 3, 62, 148, 149; heavy water (D_2O), 83; isotopes of, 82; isotopic composition of seawater, 77; isotopic forms of the water molecule, 55, 70; polar waters, 8; in the stratosphere, 170; transition between water vapor and ice, 16. *See also* oceans; water vapor
water vapor, 3–4, 16, 40, 55, 79, 115, 116, 146, 163, 170, 174–75, 176, 187, 203, 212
weather forecasting, 174–75
Weddell Sea, 26, 133
Wegener, Alfred, 28
West Africa, 44
West Antarctica, 34, 35–36, 194, 199, 213, 221; ice shelves of, 13–14
West Antarctica Ice Sheet (WAIS), 105
Western Europe, 146, 149, 180, 215, 216; rapid cooling in, 150
White, Jim, 132
Wilkes Coast, 34
wind, and the transfer of impurities, 72
World Climate Research Programme (WCRP), 177
World Conference on Climate, 177
"World Energy Outlook" (International Energy Agency [IEA]), 230, 237
World Meteorological Organization (WMO), 177, 178, 249, 251
World Research Program, Climate and Cryosphere (CLIC) initiative of, 248

Yiou, Françoise, 79, 119
Younger Dryas period, 130–31, 132, 136–37, 149; identification of in North America and Central Europe, 131

Zotikov, Igor, 255
Zuchelli, Mario, 99